Sea Change

This publication is based on research that forms part of
the Paragon Initiative.

This five-year project will provide a fundamental reassessment
of what government should – and should not – do. It will put
every area of government activity under the microscope and
analyse the failure of current policies.

The project will put forward clear and considered solutions to
the UK's problems. It will also identify the areas of government
activity that can be put back into the hands of individuals,
families, civil society, local government, charities and markets.

The Paragon Initiative will create a blueprint for a better,
freer Britain – and provide a clear vision of a new relationship
between the state and society.

SEA CHANGE

How Markets and Property Rights
Could Transform the Fishing Industry

EDITED BY RICHARD WELLINGS

with contributions from

PAUL DRAGOS ALIGICA
H. STERLING BURNETT
BIRGIR RUNOLFSSON
ION STERPAN
RACHEL TINGLE

Institute of
Economic Affairs

First published in Great Britain in 2017 by
The Institute of Economic Affairs
2 Lord North Street
Westminster
London SW1P 3LB
in association with London Publishing Partnership Ltd
www.londonpublishingpartnership.co.uk

The mission of the Institute of Economic Affairs is to improve understanding
of the fundamental institutions of a free society by analysing and expounding
the role of markets in solving economic and social problems.

A CIP catalogue record for this book is available from the British Library.

ISBN 978-0-255-36740-0

Many IEA publications are translated into languages other
than English or are reprinted. Permission to translate or to reprint
should be sought from the Director General at the address above.

Typeset in Kepler by T&T Productions Ltd
www.tandtproductions.com

Printed and bound in Great Britain by Hobbs the Printers Ltd

CONTENTS

3 The European Common Fisheries Policy   52

Rachel Tingle

4 Governing the fisheries: insights from Elinor Ostrom's work   95

Paul Dragos Aligica and Ion Sterpan

5 Rights-based ocean fishing in Iceland 117

Birgir Runolfsson

About the IEA 144

THE AUTHORS

Paul Dragos Aligica

Paul Dragos Aligica is a Senior Research Fellow at the F. A. Hayek Program for Advanced Study in Philosophy, Politics, and Economics at the Mercatus Center at George Mason University, where he teaches in the graduate programme of the Economics Department. He specialises in institutional theory, public choice and comparative economic and governance systems. He has authored seven books, including *Institutional Diversity and Political Economy: The Ostroms and Beyond* (Oxford University Press, 2014), and has written for the *Wall Street Journal Europe* and a wide variety of academic journals, including *American Political Science Review, Public Choice, Revue française d'economie* and *Comparative Strategy*. Aligica has been a consultant for the United Nations Development Program, the World Bank, European Union organisations and the United States Agency for International Development (USAID). He received his PhD in political science from Indiana University Bloomington.

H. Sterling Burnett

H. Sterling Burnett has a doctorate in applied philosophy from Bowling Green State University. He specialised in environmental ethics. He is currently a research fellow in energy and environmental issues at the Chicago-based research organisation, The Heartland Institute, and is managing editor of its lead environmental publication, *Environment & Climate News*. Before joining The Heartland Institute, Burnett served for 18 years as the senior fellow responsible for the environment programme at the National Center for Policy Analysis in Dallas, Texas.

Birgir Runolfsson

Birgir Thor Runolfsson is an Associate Professor of Economics at the University of Iceland. He has an undergraduate degree in economics from Lewis and Clark College and a masters degree and PhD in economics from George Mason University. He has published papers and edited books on fisheries management, as well as papers in the area of public choice and institutional economics. He has also participated in projects and written several reports on fisheries management issues both in Iceland and internationally.

Ion Sterpan

Ion Sterpan is an economics PhD student at George Mason University and a graduate fellow at the Mercatus Center's F. A. Hayek Program. He studied philosophy at Central European University in Budapest and at the University of Bucharest and worked in Bucharest with the Center for Institutional Analysis and Development. His research topic is polycentric institutional systems, which he approaches from an Austrian and Virginian political economy perspective.

Rachel Tingle

Rachel Tingle studied economics at the University of Exeter before becoming a research fellow at the University of York and taking a master's degree at the University of Surrey. She has had a varied career as an economist and journalist, which has included teaching at the Universities of Buckingham and Brunel, as well as working in the City, in economic consultancy, for a number of senior Conservative politicians, and in television, print and web journalism. For many years she has specialised in writing about the interface between economics, politics and Christianity and has written hundreds of articles and two books in this area.

Richard Wellings

Richard Wellings is Acting Academic and Research Director at the Institute of Economic Affairs. He was educated at Oxford and the London School of Economics, completing his PhD in 2004. He is the author, co-author or editor of several papers, books and reports, many of which focus on transport, energy and environmental policies.

FOREWORD

After a period of rapid growth from the 1950s until the 1980s, the amount of fish captured wild in the seas and oceans has levelled off in recent decades and in some regions it has declined catastrophically. By contrast, farmed fish production and agricultural yields on land continue to increase. Given the rising world population and the reliance of much of the world on sea fish for protein, this is a worrying trend. If sea fisheries continue to decline, it will cause significant hardship.

So, why is there such a difference between sea fisheries and farmed fish when it comes to production trends? The answer lies in the lack of property rights over sea fisheries. It is an example of the 'tragedy of the commons', explained by William Forster Lloyd. He described a situation in which common land was open to grazing by all. It would, of course, be over-grazed because a person would obtain the benefit of putting additional cattle on the land without bearing the cost that arises from over-grazing, which would be shared by all. In the end it would be destroyed. This is even clearer with fish stocks. For example, a trawler taking extra tuna from the ocean will benefit but the (perhaps hugely greater) cost of taking the extra tuna, in terms of lower levels of breeding, will be shared between all trawler owners over the long term.

In the case of farmland, if it is owned and there are stable institutions to protect private property, it will be used responsibly because its value depends on the present value of all the potential produce that can be obtained from the land. As St Thomas Aquinas put it, if land is not privately owned and everybody is responsible for it, nobody will take responsibility. Undefined or unenforced property rights are disastrous for environmental outcomes. This is not reasonably disputed.

It is possible, instead of using private ownership to regulate the use of property, for governments to try to regulate to ensure its responsible use. Indeed, that is precisely what happens with sea fisheries. As political economists will be aware, however, the political process does not produce the neat results of an impartial, omniscient arbitrator that the neo-classical textbooks imply. The reality of political bargaining through interest groups is that very inefficient methods of conserving fish stocks are used (if any at all) and, to make matters worse, trawler owners are actually subsidised, thus reducing the cost of capital in the industry.

Of course, until recent times, property rights in the sea were unnecessary. Private property is only necessary to deal with scarcity. When populations were lower and technology less advanced, scarcity was only a local problem.

For this reason, and perhaps because of a lack of imagination in relation to some technical problems, property rights in sea fisheries have not tended to develop. Of course, philosophically, the environment is not conducive either. In the era when property rights solutions have become needed, governments have tended to look to regulation instead. However, there are examples around the world of regimes that provide long-term property rights interests in fish being very successful. Exactly how these work varies from place to place depending on the institutional setting and the specific features of different fisheries.

The authors of this important book explain, in various different contexts, how private property rights systems can be used to ensure flourishing fisheries. Strong institutions are important and economic analysis is key for defining the limited role that government should play.

There is much discussion of the future of fisheries policy in the UK, especially given its relevance post-Brexit. The principles are clear from the work of the authors of this book. There needs to be an institutional framework so that the owners of fishing rights become enthusiastic conservationists, as has happened in countries such as Iceland. The only point of reasonable dispute is how we deal with the practicalities. Again, the authors provide useful guidance on this.

Another theme of this book is rarely discussed in other debates on fishing. The political economy pressures don't only lead to poor decisions when it comes to how to regulate fishing, they positively encourage investment in the industry thus encouraging overcapacity. The editor, Richard Wellings, lists as many as five 'do no harm' actions that the government could take which involve the reversal of artificial encouragement towards intensive fishing. Governments should take note.

Finally, it is worth noting that effective regimes of private property rights are important for resolving conflicts (this was another point made by Aquinas). There is, at the current time, considerable concern about the impact of aquaculture on sea fisheries. Problems might arise as a result of the escape of parasites or of fish with particular genetic characteristics. Indeed, these problems could be very serious indeed. If there were property rights in sea fisheries, courts could require compensation for losses if the action of fish farmers damaged sea fisheries. Fish farmers would then be incentivised to use secure methods of farming (which do exist) and would not have to be told to do so by government regulation.

Property rights solve problems of scarcity, they promote conservation, they assign responsibilities and they promote the peaceful resolution of potential conflicts. It is to be hoped that politicians and their advisers and all who influence debates on this subject will understand the importance of this book.

The views expressed in this monograph are, as in all IEA publications, those of the authors and not those of the Institute (which has no corporate view), its managing trustees, Academic Advisory Council members or senior staff. With some exceptions, such as with the publication of lectures, all IEA monographs are blind peer-reviewed by at least two academics or researchers who are experts in the field.

Philip Booth

Professor of Finance, Public Policy and Ethics and Director of Research and Public Engagement at St Mary's University, Twickenham, and Senior Academic Fellow at the Institute of Economic Affairs

February 2017

SUMMARY

- Global fish catches in the seas and oceans have stagnated since the mid 1990s. A decline in the quality of fish landed has been evident in several major regions and some fisheries have experienced collapses in stocks of valuable species such as cod.

- Because there are generally no established property rights in wild fish, fisheries are vulnerable to the 'tragedy of the commons'. Trawler owners race to catch as many fish as possible before they are caught by competitors. While the short-term benefits of overfishing accrue to the individual trawler owners, the long-term costs in terms of reduced yields are shared.

- Governments have greatly exacerbated the problem of overfishing by subsidising the industry and undermining the economic feedback mechanisms that help to protect stocks. State support has kept commercially loss-making fishing enterprises in business, creating significant overcapacity.

- Counter-productive subsidies reflect the influence of the fishing industry over government policy. Tightly knit groups of trawler owners and fishermen have strong incentives to lobby governments for financial support, whereas individual taxpayers have weak incentives to lobby against it.

- The European Union's Common Fisheries Policy (CFP) has been particularly prone to political influence, with disastrous results. The British fishing industry, for example, has been in almost continuous decline in recent decades as stocks have fallen. Landings into UK ports of the more valuable demersal fish such as cod have plummeted by around 80 per cent since 1970. The UK shares fishing grounds with other member states and has been allocated a relatively small share of EU quotas.
- Under the CFP, a high proportion of fish caught have been thrown back dead into the sea. For the period 2003–5, discard rates within EU waters were running at 20–60 per cent of the catch weight for typical fisheries exploiting demersal fish. Between 1990 and 2000, over 500,000 tonnes of fish were discarded annually just in the North Sea.
- Brexit will enable the UK to withdraw from the CFP and adopt a more efficient approach within its large Exclusive Economic Zone, which stretches up to 200 nautical miles from the coast. This has the potential both to increase catches and eliminate subsidies from taxpayers. Policy options include facilitating community-based management in some coastal fisheries and introducing Individual Transferable Quotas (ITQs) for other areas.
- Many local communities around the world have evolved successful approaches to managing coastal fisheries without the need for government intervention. They set their own rules on who has access to the

resource, how it can be fished and what sanctions will be imposed if violations occur. Such management models have typically been highly effective at conserving stocks and maintaining yields in the long term, in marked contrast to the failure so often observed under state regulation.

- Property-rights-based systems, such as Individual Transferable Quotas, alter incentives in ways that are favourable to conservation and stewardship, because overfishing reduces the value of the quotas owned by fishermen. Where they have been introduced, ITQs have improved efficiency by reducing excessive fishing effort and over-capitalisation.
- Although the success or failure of any policy approach will be affected by the characteristics of the fishery to which it is applied, there are general lessons. An economically rational strategy for UK waters post-Brexit, and other fisheries currently subject to mismanagement, will necessarily involve phasing out government subsidies in all their forms and better aligning incentives with the long-term preservation of stocks.

TABLES AND FIGURES

1 INTRODUCTION

Richard Wellings

The depletion of fish stocks has proved to be a difficult problem to resolve. A high proportion of the world's fisheries are thought to be 'over-exploited' and yields are likely to fall in the long term. Declining catches will then have a significant wider economic impact. Around 35 million people are directly employed by the fishing industry, with a multiple of this engaged in processing and support jobs (FAO 2014). Moreover, fish comprise roughly 20 per cent of the animal protein in people's diets and often a much higher percentage in poorer countries (ibid.). While the fishing industry makes up only a small fraction of 'gross world product', perhaps 0.25 per cent, its share of GDP is far larger in many developing economies, including Indonesia, the Philippines and Vietnam.[1]

The importance of the 'fisheries crisis' goes beyond its economic impact. The depletion of stocks is symptomatic of a far wider problem with 'open access' resources. Policies that prove effective for the fishing industry are therefore

---

1   See, for example, 'Contributions of fisheries and aquaculture in Asia and the Pacific region', FAO website (http://www.fao.org/docrep/011/i0433e/I0433E04.htm).

likely to have wider relevance to issues such as overgrazing of grasslands and deforestation – at least in terms of general principles.

This collection examines the economic and political causes of the problems faced by fisheries and also sets out potential solutions. If a theme unites the contributions, it is that one-size-fits-all approaches are ill-suited to this policy area, both in analysing how depletion arises and developing strategies to address it. Moreover, measures that might appear attractive in terms of economic theory may be undermined by the incentives facing political actors in the locations where they are applied.

The remainder of this introductory chapter provides a critical overview of the major issues facing the fisheries sector and a summary of key theoretical explanations for the current crisis. It argues that standard analyses of 'overfishing' are often simplistic and fail to acknowledge adequately the role of states in undermining market feedback mechanisms that would help to protect stocks.

Similarly, government action often weakens or destroys voluntary and/or community-based arrangements between fishermen, sometimes with disastrous consequences. Finally, it is clear that limited intervention in the form of the creation and enforcement of property rights is likely to be more efficient than more active state regulation of the sector.

Global fish stocks

The foregoing discussion alludes to severe problems with the world's fisheries. Indeed, much analysis of this issue begins with the assumption that several species are close to depletion. It is claimed that the oceans are likely to become 'virtual deserts' within the next few decades.[2] The reality is rather more complex. In some respects the fishing industry could be viewed as a success story. Overall global fish 'production' has increased by a factor of eight since 1950, far outstripping population growth and enabling per capita consumption of this protein-rich food to almost treble to 19 kg per year (FAO 2014).

The aggregate data hide some worrying trends, however. Following a long period of steady growth, annual catches of wild fish in the seas, oceans and inland waters appear to have stagnated at around 90 million tonnes since the mid 1990s (Figure 1). A major expansion in fish farming therefore explains recent increases in supply (Figure 2). In particular, market-friendly reforms in China since the late 1970s have encouraged rapid growth in its aquaculture sector (FAO n.d.).

In the same period, many long-established sea fisheries have experienced substantial declines in catches. In the North West Atlantic, for example, landings have fallen by approximately 55 per cent since the 1968 peak (FAO 2011). Worse still, within these totals, the yield of certain species has collapsed. Perhaps best known is the 98 per cent fall in cod catches in the North West Atlantic between 1968 and 2003 (ibid.: 24), with the species virtually disappearing from the seas off Canada where it was once abundant.

---

2 As an illustrative example, see 'The end of fish: the ones we like to eat are rapidly vanishing from the ocean', *Washington Post*, 3 June 2014 (https://www.washingtonpost.com/posteverything/wp/2014/06/03/the-end-of-fish/).

Figure 1   Global fish catch, 1950–2013 (wild capture)

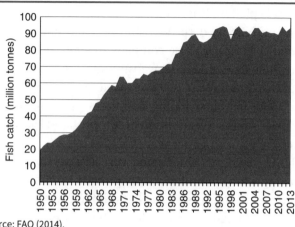

Source: FAO (2014).

A decline in the quality of catches has also been evident in several major regions. The share of high-value fish such as cod and wild salmon has been falling, while that of low-value species such as blue whiting and sand eels – typically used to make fishmeal or fish oils – has increased. Aggregate tonnage figures may therefore mask the extent to which fish stocks have been degraded. Moreover, they may not reflect the environmental damage caused, for example, to seabeds by beam trawling[3] or to marine creatures not targeted by fishermen but nonetheless killed or injured by their practices (see Moore and Jennings 2000).

---

3   In beam trawling, a large net is attached to a heavy metal beam which is dragged across the seabed behind a trawler, often destroying marine animals and their habitats in the process.

Figure 2 Global fish production from aquaculture, 1950–2013

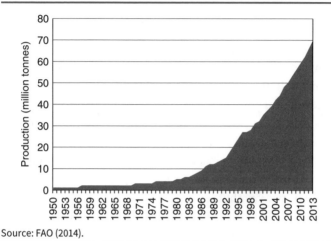

Source: FAO (2014).

Given the decline in many long-established fisheries, growing catches in previously little-exploited regions such as the Eastern Indian Ocean have helped maintain levels of global supply (ibid.). Nevertheless, there are fears that these emerging fisheries will follow a similar trajectory, with growth phases a precursor to declining overall yields and a collapse in high-value species.

Such a hypothesis underlies the predictions that global fish catches will fall precipitously by the middle of the twenty-first century, with this view often combined with forecasts that climate change and marine pollution will also have a devastating impact on yields (Worm et al. 2006; Allison et al. 2009).

There is good reason to believe, however, that such concerns are overstated. Firstly, in many fishing grounds stocks appear to be stabilising or even recovering after previous slumps. And, within such fisheries, the situation may vary markedly by species. This makes an across-the-board collapse unlikely. Secondly, at least some of the environmental problems seem to have been mitigated, particularly in developed regions where high living standards have increased the demand for 'environmental goods' such as clean seas and fish caught without damaging other species. Thirdly, there is arguably growing awareness of the policy mistakes that led to the depletion of fisheries in the past. Finally, the growth of aquaculture, or fish farming, may enable the production of fish in both developed and developing countries at costs that are lower than fishing in the open seas in a situation of declining stocks.

All these issues are intertwined. However, it is the failures of past government policy and how they have contributed to current problems that is a major theme of this monograph. In order to analyse the most important theoretical issues and policy questions, the discussion focuses on sea fisheries rather than rivers and lakes, and on the continental shelf and inshore resources more than the relatively barren high seas.

The tragedy of the commons

The concept of the 'invisible hand' was developed by Adam Smith, who explained how an individual who 'intends only his own gain' may be 'led by an invisible hand to promote...

the public interest'. Unfortunately, Smith provides very few examples of situations where this principle would and would not work, or criteria to explain this. However, in the case of resources such as fisheries, the notion has been heavily criticised, though perhaps unfairly.[4]

Most famously, Hardin (1968) describes the 'tragedy of the commons'[5] in which individuals acting in their own self-interest bring ruin to all. He uses the example of a pasture open to anyone who wishes to graze his animals there (stating later that the problem also applies to the oceans (ibid.: 1245)):

As a rational being, each herdsman seeks to maximize his gain. Explicitly or implicitly, more or less consciously, he asks, 'What is the utility to me of adding one more animal to my herd?' ... Since ... the effects of overgrazing are shared by all the herdsmen ... the rational herdsman concludes that the only sensible course for him to pursue is to add another animal to his herd. And another ... But this is the conclusion reached by each and every rational herdsman sharing a commons. Therein is the tragedy. Each man is locked into a system that compels him to

4 Arguably this critique represents a narrow interpretation of Smith's concept, which might be considered as embedded in his wider ideas about political economy, including the importance of institutions such as private property rights.

5 Although the tragedy of the commons is generally attributed to Hardin, he is actually referring back to a pamphlet by William Forster-Lloyd, from whom the idea probably originates. Hardin is using the idea of the agricultural commons to explain his proposals for authoritarian control of population growth.

increase his herd without limit – in a world that is limited. Ruin is the destination toward which all men rush, each pursuing his own best interest in a society that believes in the freedom of the commons.'

According to this view, the only way to address the damaging effects of such behaviour is to have an external body impose a management structure over the resource – whether in the form of private ownership, government ownership, or state regulation (see Pennington 2012: 23).

This perspective has been challenged by researchers such as Elinor Ostrom, as set out in Chapter 4. In many instances, communities of resource users have managed to evolve effective rules and management structures that have avoided the tragedy of the commons without government involvement. Indeed, this is frequently the case for inshore fisheries, where cooperatives and associations prevent over-exploitation through, for example, limits on individual catches. In response to such evidence, Hardin later suggested that a more accurate title for his original paper would have been 'The tragedy of the unmanaged commons' (Hardin 1994).

Hardin's scenario is perhaps more applicable to the open seas than to inshore waters, which may be subject to effective community-based management. Fish stocks in the high seas, for example, are theoretically open to all and trawlers from all over the world are free to exploit them. Under such circumstances there are major obstacles to the evolution of the kind of community-based rules seen in many inshore fisheries.

Open access also applied to large areas of the shallow, fish-rich continental shelf before the imposition of Exclusive Economic Zones in the second half of the twentieth century. These eventually extended up to 200 nautical miles from coastlines and effectively granted the relevant governments ownership of the fish and mineral resources within them. Until the mid 1970s, for example, British trawlers caught substantial tonnages above the continental shelf around Iceland (Gissurarson 2000: 12). From Hardin's perspective there should be a strong case for governments to impose management structures on open-access fisheries, whether through direct regulation or the award of property rights to fishermen.

How market mechanisms protect fisheries

Some theoretical objections should be noted, however. In the case of unmanaged open-access fisheries, incentive structures would appear to be far more complex than suggested in Hardin's classic example. In particular, there are typically strong negative feedback mechanisms that reduce the likelihood of 'ruin', such as the catastrophic reduction in yields seen in North West Atlantic cod.

As stocks begin to decline, this will tend to affect negatively the viability of fishing enterprises, with marginal operators going out of business as costs increase and revenues decrease (see Gordon 1954; Clark 1990). The precise impact will of course depend on the impact of increased scarcity in a particular fishery on fish prices. Under conditions of free trade the effect on prices is likely to be diluted as supplies can be sourced from other regions that are not experiencing depletion. Scarcity-driven price increases for marine fish will also tend to encourage investment in alternative forms of production such as fish-farming.

Market feedback mechanisms will also figure in the calculations of fishing entrepreneurs such as trawler owners, who will be alert to market conditions. They must consider the behaviour of their competitors. If other businesses are more effective at capturing the reduced stocks, this will change the point at which the costs of additional fishing activity outweigh the benefits. In situations where search costs increase, economies of scale decline and so on, but prices do not rise sufficiently to compensate for this, entrepreneurs at the margin will tend to withdraw from the market. It should be noted that, unlike some grazing scenarios, the marginal costs of fishing activity may be rather high, including crewing costs, fuel, and wear and tear to equipment (see Abernethy et al. 2010).

This is not to argue that there cannot be circumstances in which a certain species gains 'scarcity value' as a rare delicacy or luxury item, such that prices rise sufficiently to cover the increased costs associated with falling stocks, leading to a 'death spiral' for the fishery (Courchamp et al. 2006). This would appear to be a special case, however, and does not explain the major collapses of mass-market species observed in recent history. Purported empirical examples, such as the Napoleon wrasse (a reef fish now extinct in many areas), are questionable due to their juxtaposition with industry subsidies.

Consumers may also choose to respond to price rises by substitution, for example, eating pollock instead of cod. And they may change their behaviour due to concerns about the impact they might be having on fish stocks and the environment more generally – assuming that media and campaign groups are at liberty to raise awareness of such problems. Robinson (2008: 64) explains how such consumer pressure could mitigate potential climate change problems, but his analysis could apply equally well to the conservation of fisheries:

> If there is general concern that the natural environment is becoming overused, the effect may be as if it were owned. Actions by individuals are characterised not so much by narrow self-interest (in the self-centred sense) but by broader interests that include concern for family, friends and descendants. Let us assume that a large part of the population is very concerned about the world in which their children and grandchildren will grow up. In such circumstances, one would expect that both consumers and producers (the latter both spontaneously and as a reaction to the views of consumers) would act in ways they perceive would protect their successors ... [and] will demand and will be supplied with goods and services that are deemed 'green' ... The 'perpetual referendum' that constitutes the market – which means that people are voting every day by expressing their preferences – will produce votes for 'green' outcomes which producers, in their own self-interest ... will satisfy.

This explains why producers and retailers have adopted environmentally friendly practices and labelled their goods accordingly. Examples include 'dolphin friendly' tuna, 'pole and line' caught tuna and 'certified sustainable' seafood.[6] Another initiative encourages the public to boycott restaurants that serve bluefin tuna, a species at risk of collapse.[7] Only a fraction of fish consumers need to adopt such purchasing habits for such standards to be widely adopted in the sector, given issues of branding and economies of scale. Producers and retailers may also decide to foster sustainable fisheries practices in order to avoid risks of shaming and reputational damage.

Nevertheless, we would not expect these mechanisms to entirely mitigate the tendency towards overfishing. The benefits of behaviour that promotes conservation may be shared more widely than the costs incurred by those individuals who choose the more prudent course. There is, though, a further factor at work that encourages overfishing.

Cutting off the invisible hand

Hardin (1968: 1246) explicitly acknowledges the role of negative feedback in addressing the tragedy of the commons in his discussion of what he perceives as a problem of overpopulation:

---

6 The latter scheme is operated by the Marine Stewardship Council. See https://msc.org/ for more details.

7 See, for example, http://www.biologicaldiversity.org/species/fish/Atlantic_bluefin_tuna/bluefin_boycott/

In a world governed solely by the principle of 'dog eat dog' ... how many children a family had would not be a matter of public concern. Parents who bred too exuberantly would leave fewer descendants, not more, because they would be unable to care adequately for their children.

However, as he goes on to explain, modern society is deeply committed to the welfare state, which means the full costs of reproductive choices are not experienced by the individuals making the decisions.

In a similar way, the market mechanisms that would typically offer fish stocks some protection from catastrophic decline are undermined by the welfare state for the fishing industry. Governments around the world pump vast subsidies into the sector (see Chapter 2) which encourage overfishing by increasing the size of the fishing fleet and blunting the market processes that would help to preserve stocks. This welfare state effectively shields fishing entrepreneurs from the negative consequences of bad decisions, and encourages investment that would otherwise be seen as too risky.

The fishing industry is further distorted by various indirect interventions. For example, welfare benefits themselves may deter fishermen from seeking other employment and/or moving location by providing income during lean periods (see Tietze 2016). Governments also fund the construction of harbour facilities and subsidise transport links that connect fishing ports to larger markets. This taxpayer-funded infrastructure often would not have been commercially viable. Access to some fishing grounds – for

example, off the coasts of many less developed countries
– may be contingent on state spending on defence, foreign
aid, diplomacy and so on. Such interventions will once
again tend to lower industry costs and as a consequence
contribute to overfishing. This is not to neglect those state
policies that raise costs, such as taxation and regulation,
though it is notable that in many countries the fishing
industry is particularly heavily subsidised compared with
other economic sectors. According to one estimate, direct
subsidies alone are equivalent to roughly 25 per cent of the
value of catches worldwide (UNEP 2008).

Fishing for favours

Economic incentives provide a plausible explanation for
the political favours awarded to the industry. Concen-
trated interests such as the fishing sector have very strong
incentives to deploy resources on lobbying policymakers
since the potential payoffs are large (see Olson 1965). Examples are manifold, including aggressive protests by fishermen who have blockaded commercial ports in attempts
to pressure their governments to meet their demands.[8]
Such activity is further facilitated by the strong social
bonds typical of fishing communities, which help to prevent free-riding, i.e. individuals letting other people do the

---

8   For example, 'Scottish fishermen blockade three harbours in quota row',
    The Independent, 1 June 1993; 'Spain deal to end fish blockades', *BBC News*,
    27 October 2005; 'French fishermen resume Channel blockade', *Daily Tele-
    graph*, 16 April 2009.

work of campaigning but still enjoying the benefits of the action of other campaigners.

By contrast, the losers from fishing subsidies are typically general taxpayers, a very large and highly dispersed group in which individual losses are small and barely noticeable. Moreover, their social bonds are weak, making organising political action difficult. The incentives are therefore poor for any sort of collective action opposing government funding of the fishing industry. Other losers include, of course, future generations of consumers and trawler owners who will be faced with reduced fish stocks. Again, it is difficult – if not impossible – for such groups to coordinate

While the 'logic of collective action' provides a plausible explanation for subsidy regimes, it should be noted that several policy rationales may be given. For example, food security has been used to justify state support for the large fleets of Japan and Taiwan, both mountainous countries with relatively little arable land (FAO 2000). Subsidies may also be channelled to the fishing industry through regional policies that attempt to reduce geographical inequalities by boosting the economies of struggling peripheral areas, as has happened in the European Union (see Chapter 3).

The tragedy of state regulation

As set out in Chapter 2, there is a great deal of evidence that subsidies are largely to blame for the declining fish stocks observed in many regions. The strength of the fishing lobby has, however, made it difficult for policymakers to withdraw them. Solutions have therefore centred on regulating the industry in order to limit catches, but at the same time continuing the subsidies that shield fishing businesses from the consequences of their behaviour. While this approach is illogical and contradictory from an economic perspective, it would appear to be rational in terms of the political incentives described above.

Regulation is not costless. Like subsidy regimes, it is typically captured by special interests (Stigler 1971) and subject to political interference. In addition, the bureaucrats employed to oversee the regulatory system may have incentives to increase the scope and scale of their intervention in order to increase their salaries, job security and status (Niskanen 1971; Dunleavy 1991). Regulators also face severe problems gathering the information required to allocate resources efficiently – in setting fish quotas, for example. It may be difficult for them to access the constantly changing, time and place specific knowledge relevant to their decisions (see Hayek 1945). Moreover, any control system will involve transaction costs such as complying with rules and paying for administration and enforcement.

These are key reasons why the costs of introducing regulations may exceed the benefits. Accordingly, even if market failure is perceived and unregulated markets generate outcomes that are viewed as suboptimal, this does not necessarily justify state intervention.[9]

9　The methodological problems associated with quantifying costs and benefits should also be noted, as well as the incentives for cost–benefit analyses to be manipulated to favour particular outcomes.

This conclusion is well illustrated by the disastrous history of interventions in the fishing industry. The European Union's Common Fisheries Policy, a complex system of centrally imposed quotas, regulation and subsidies examined in Chapter 3, is a case in point. Under EU control many fish stocks, particularly relatively valuable demersal[10] species such as cod, fell below safe biological limits, though there has been some recovery of late. Indeed, UK landings of these fish were about 80 per cent lower in 2014 than in 1970 (before the UK joined the EEC). In many fishing ports, this endemic mismanagement has resulted in large-scale job losses and economic decline.

Yet the political economy of fisheries means that a laissez-faire approach is unlikely to be adopted. So long as direct and indirect subsidies continue, countervailing regulation will be required to address the harmful consequences. Moreover, even if an individual government adopts a strict no-subsidy policy, spillover effects from the actions of other states may encourage some form of intervention in response. If heavily subsidised industrial trawlers from, say, Japan, Taiwan or the EU are stripping a local resource, the case for some means of excluding them is strengthened. Even without subsidy, there might be situations where overfishing will occur. The key question then becomes the effectiveness of different regulatory approaches in maximising the economic returns from fisheries and achieving other policy objectives.[11]

---

10 Demersal species live on or near the seabed.

11 Other objectives may include the protection of marine environments, etc.

Property rights approaches

Within government-managed zones it would appear that some regulatory approaches are more effective at delivering economic benefits than others. In particular, there is evidence that approaches based on property rights are more successful at meeting key objectives than more active regulation and micromanagement of the industry.

As explained in Chapter 5, Individual Transferable Quotas (ITQs) grant fishing businesses a share of the total catch in a given fishery, reducing incentives for overfishing. The overall catch for each species is set by a government regulator, informed by data on fish stocks. Importantly, the quotas can be bought and sold, providing financial incentives for less-efficient businesses to sell their rights to more efficient ones. ITQs can also avoid many of the gross inefficiencies characteristic of other regulatory strategies such as the imposition of short fishing seasons to limit catches or prohibitions on certain types of equipment.

An effective ITQ system should give the rights to trawler owners to a given share of the total catch in perpetuity (or, at least, provide sufficient legal ambiguity that this is the understood position, as is the case in Iceland). This means that trawler owners have an incentive not to overfish as they are the beneficiaries of conservation due to their rights stretching out into the indefinite future. In such systems, the right to establish the total allowable catch each year can, in fact, be given to the trawler owners. Thus the trawler owners become conservationists and fundamental conflicts within the system are reduced.

Yet the development of such property rights is no panacea for fisheries management. It remains vulnerable to special interest influence and politicisation. Moreover, if there is government central planning of total allowable catches, this may suffer from knowledge limitations, in the sense that decisions may not draw on the changing, time- and place-specific information available to fishing businesses.[12] And the operation of the system may involve substantial transaction costs.

A more principled criticism is that the implementation of both Exclusive Economic Zones and ITQ systems involves state aggression, arguably exemplified in the former case by the Cod Wars,[13] together with a significant degree of government discretion in the initial allocations and system design. In the Cod Wars, the Icelandic government effectively discriminated against foreign boats, thus destroying their livelihoods. Furthermore, vast marine territories may be grabbed through nominal and illegitimate state 'ownership' of small, uninhabited islands. Boundary disputes, as observed today in the South China Sea, have the potential to escalate into major conflicts between governments.

Injustices may be a particular problem in developing countries. In many cases, artificial nations were created

---

12 Although, as noted above, conceivably a model could be adopted whereby catches were set by the industry itself, given the strong economic incentives for maintaining stocks (see Booth 2016).

13 During the Cod Wars, the Icelandic government extended the exclusive fishing zone around its coastline and forced out British trawlers that had traditionally fished these grounds.

by external powers, often with little regard for natural boundaries,[14] and with a top-down rule of law determined by elites supplanting traditional and widely respected local practices. Corrupt government officials may also have strong incentives to rig regulatory structures in favour of industrial fishing at the expense of small-scale, semi-subsistence based activity – the former providing substantial licensing income that can be siphoned off by state officials or used to pay off special interests (see Standing 2008).

Such difficulties raise the question of whether voluntary strategies could be applied to fisheries. For example, it has been suggested that areas of sea could be transferred to private ownership through a process of 'homesteading' (McElroy 2014). The 17th-century philosopher John Locke explained how unowned land could be acquired by individuals who 'mixed their labour' with it, for example, by clearing forest, ploughing the soil and planting crops. Yet it is not clear how this principle should be applied to hunter-gatherer societies such as Native American tribes, who did not mix their labour with the extensive hunting grounds they depended on (see DiLorenzo 1998).[15] Similarly, the marine fishing industry does not typically mix its labour with the oceans; rather fish stocks develop naturally before being gathered by fishermen.

---

14 For example, based on physical geography and/or ethno-linguistic divisions.

15 Although they did modify the land through use of fire, to promote grass-lands rather than forest.

Conclusions

Economic theory suggests that state intervention in the fishing sector will create major problems due to politicisation, special interest influence, knowledge limitations and the imposition of transaction costs. These conclusions are amply illustrated by the disastrous history of government involvement, with many fisheries depleted, substantial environmental damage and taxpayers forced to subsidise the industry.

Ideally, then, the primary objective of fisheries policy should be to remove harmful interventions. An economically rational approach might include the following components, which in the UK context could provide a broad framework for the government's strategy post-Brexit:

- Phasing out direct subsidies to the fishing industry.
- Removing indirect subsidies such as grants for harbour improvements and transport links to fishing ports. Such infrastructure should be operated on a commercial basis to ensure that capital and maintenance costs are reflected in the charges paid by fishing businesses.
- Reforming welfare policies and removing barriers to labour mobility in order to increase the opportunity cost of staying in the fishing industry rather than seeking better-paid employment elsewhere.
- Allowing free trade in fish products to reduce the risk of species developing 'scarcity value' in a particular region.

- Reducing regulatory barriers to aquaculture to enable consumers to substitute depleted species for similar farmed fish, while at the same time removing subsidies for fish farming that may encourage overfishing of species used as feed.
- Refraining from imposing state regulation on inshore fisheries already effectively managed by community-based rules.
- Encouraging other states to end harmful interventions, for example through diplomacy, voluntary boycotts or perhaps as a component of trade agreements.
- Allowing freedom of expression and competition such that consumer pressure is able to encourage the adoption of more sustainable practices.
- Where state action is difficult to avoid (perhaps due to the impact of subsidies by other governments), limiting intervention to the creation and enforcement of property rights, such as Individual Transferable Quotas.

While some of these solutions are acknowledged within the policy community, the extent to which state intervention contributes to overfishing is not widely appreciated. The assumption that depletion is an inevitable result of market failure still holds considerable sway. Yet even if the intellectual case for a free-market approach to fisheries were accepted, the fishing industry would almost certainly retain its disproportionate influence over the policymaking process, making fundamental reform difficult.

Granting property rights to the industry is one way of aligning incentives to improve outcomes, although this does not solve all of the problems associated with government control. Bottom-up, community-based solutions tend to be an effective option for many inshore waters. Their rules are typically closely aligned with the interests of local fishermen, but their geographical applicability is limited and they will often be vulnerable to political interference.

Fisheries issues are inevitably caught up in wider debates about the proper role of government. The industry is always likely to be subject to heavy regulation in an era when most economic sectors are under relatively tight bureaucratic control. Lasting solutions must therefore be accompanied by a shift in wider political culture. Granting special privileges to the fishing industry would then be viewed as outside the legitimate remit of the state.

References

Abernethy, K. E., Trebilcock, P., Kebede, B., Allison, E. H. and Dulvy, N. K. (2010) Fuelling the decline in UK fishing communities? *ICES Journal of Marine Science* 67: 1076–85.

Allison, E. H, Perry, A. L., Badjeck, M.-C., Adger, W. N., Brown, K., Conway, D., Halls, A. S., Pilling, G. M., Reynolds, J. D., Andrew, N. L. and Dulvy, N. K. (2009) Vulnerability of national economies to the impacts of climate change on fisheries. *Fish and Fisheries* 10(2): 173–96.

Booth, P. (2016) *A Briefing: Fisheries Policy outside the EU.* London: Institute of Economic Affairs.

Clark, C. W. (1990) *Mathematical Bioeconomics: Optimal Management of Renewable Resources*. Hoboken, NJ: Wiley.

Courchamp, F., Angulo, E., Rivalan, P., Hall, R. J., Signoret, L. and Bull, L. (2006) Rarity value and species extinction: the anthropogenic Allee effect. *PLoS Biology* 4(12): e415.

DiLorenzo, T. (1998) The Feds versus the Indians. *The Free Market*, 16(1).

Dunleavy, P. (1991) *Democracy, Bureaucracy and Public Choice*. London: Pearson Education.

FAO (2000) *Sustainable Contribution of Fisheries to Food Security*. Rome: FAO.

FAO (2011) *Review of the State of World Marine Fishery Resources*. Rome: FAO.

FAO (2014) *Fishery and Aquaculture Statistics*. Rome: FAO.

FAO (n.d.) *National Aquaculture Overview: China*. Rome: FAO (http://www.fao.org/fishery/countrysector/naso_china/en).

Gissurarson, H. H. (2000) *Overfishing: The Icelandic Solution*. London: Institute of Economic Affairs.

Gordon, H. S. (1954) The economic theory of a common property resource: the fishery. *Journal of Political Economy* 62: 124–42.

Hardin, G. (1968) The tragedy of the commons. *Science* 162: 1243–48.

Hardin, G. (1994) The tragedy of the unmanaged commons. *Trends in Ecology and Evolution* 9(5): 199.

Hayek, F. A. (1945) The use of knowledge in society. *American Economic Review* 35(4): 519–30.

McElroy, W. (2014) Obama wants to close the oceans. Privatize instead! Future of Freedom Foundation, 17 July (http://www.fff.org/explore-freedom/article/obama-wants-to-close-the-oceans-privatize-instead/).

Moore, G. and Jennings, S. (eds) (2000) *Commercial Fishing: The Wider Ecological Impacts.* London: British Ecological Society.

Niskanen, W. A. (1971) *Bureaucracy and Representative Government.* Chicago and New York: Aldine Atherton.

Olson, M. (1965) *The Logic of Collective Action: Public Goods and the Theory of Groups.* Cambridge, MA: Harvard University Press.

Robinson, C. (ed.) (2008) *Climate Change Policy: Challenging the Activists.* London: Institute of Economic Affairs.

Pennington, M. (2012) Elinor Ostrom, common-pool resources and the classical liberal tradition. In *The Future of the Commons* (ed. E. Ostrom). London: Institute of Economic Affairs.

Scott Gordon, H. (1954) The economic theory of a common-property resource: the fishery. *Journal of Political Economy* 62(2): 124–42.

Standing, A. (2008) *Corruption and Industrial Fishing in Africa.* Bergen: Anti-Corruption Resource Centre.

Stigler, G. (1971) The theory of economic regulation. *Bell Journal of Economics and Management Science* 2(1): 3–18.

Tietze, U. (2016) *Technical and Socioeconomic Characteristics of Small-Scale Coastal Fishing Communities, and Opportunities for Poverty Alleviation and Empowerment.* Rome: FAO.

UNEP (2008) *Fisheries Subsidies: A Critical Issue for Trade and Sustainable Development at the WTO.* Geneva: United Nations Environment Programme.

Worm, B., Barbier, E. B., Beaumont, N., Duffy, J. E., Folke, C., Halpern, B. S., Jackson, J., Lotzel, H. K., Micheli, F., Palumbi, S. R., Sala, E., Selkoe, K. A., Stachowicz, J. J. and Watson, R. (2006) Impacts of biodiversity loss on ocean ecosystem services. *Science* 314: 787–90.

2 SUBSIDISING DECLINE: GOVERNMENT INTERVENTION IN THE FISHING INDUSTRY

H. Sterling Burnett

For centuries, North America's oceans have ranked among the most bountiful on the planet. Five hundred years ago, the English explorer John Cabot reported that the waters off Newfoundland were so thick with cod that you could catch them by hanging baskets over the ship's side (Norcliffe 1999). In the waters of the Chesapeake, oyster beds were so thick they posed navigational hazards for ships. As late as the 1980s, the US – together with many other parts of the world – entered into a fisheries boom, as new fish stocks were discovered and fishing fleets expanded.

Today, US waters contain 956 fish stocks. Not counting subsistence fishing, US commercial and recreational saltwater fishing combined generated more than $199 billion in sales and supported 1.7 million jobs in 2012 (NOAA 2014). This seems substantial, but in recent years American and many world fisheries have entered a period of rapid and unprecedented decline (Hampton et al. 2005):

- A study published in the journal *Nature* notes that, in the past fifty years, populations of large fish species

- – including tuna, swordfish, marlin, sharks, cod,
  halibut and flounder – have decreased by 90 per cent.
- In US waters, the fisheries containing these same
  species have been reduced to 10 per cent of their
  historic levels (Hampton et al. 2005).
- Altogether, the National Marine Fisheries Service
  (NMFS) lists 68 species of fish as overfished (NOAA
  2013).
- Unfortunately, many commercially valuable,
  recognisable fish species are still threatened by
  overharvest in all or parts of their range, including
  multiple subspecies of marlin, tuna, cod, haddock, red
  snapper, grouper, flounder, sailfish and salmon.
- To take one prominent example, by 2004, Atlantic cod
  (once so abundant in American waters that they were
  called the 'beef of the sea') were fished to the verge of
  commercial extinction (Hutchings and Reynolds 2004).

While a number of fish stocks have recovered or are recovering, especially in the US, worldwide more than 30 per cent of fish stocks continue to be overfished. The number is even higher for commercially valuable top-predator species, which experienced a decline of more than two thirds during the twentieth century – with most of that decline occurring since the 1970s (Shiffman 2014). The UN Food and Agriculture Organization (FAO) paints an even bleaker picture. It estimates that 85 per cent of the world's fisheries are either overexploited, fully exploited, depleted, in decline or recovering from overexploitation (Oceana 2011). In Europe, for example, an estimated 63 per cent of the fish stocks examined in the Atlantic and 82 per cent in the Mediterranean are overfished (ibid.).

Why have the fisheries declined?

The decline is a result of two main factors: the institutional structure of the fisheries and misguided government policies. Concerning the institutional structure: unlike cattle, sheep and horses, fish in the ocean do not have owners. They are common property – to which everyone has access. Because they have no owners, they have no protectors or defenders. As explained at greater length in Chapter 1, the result can be what economists call the 'tragedy of the commons' (Hardin 1968).

In fisheries, all things are not equal, however. Until recently, the fact that fisheries were commons was not a problem because commercial fishing is expensive and resource intensive, while the fish stock waxes and wanes. Misguided government policies overcame the natural limits to the high cost of fishing in the commons and thus bear a large share of the responsibility for the decline of the world's fisheries.

Although most fish species can sustain occasional overfishing, prolonged periods of population loss can be critical and can lead to a collapse, when a species undergoes an abrupt, severe, sustained decline from which stocks may not recover (Walters 2005). The key to managing fish – like any other renewable resource – is therefore to implement policies that encourage fishermen to take fish in numbers that will provide for human consumption without outstripping the species' ability to reproduce itself.

Historically, allowing the fisheries to be treated as an open or anarchic commons has not been a problem, and it may still be the best policy for most ocean resources, most of the time. There are many species for which market demand is low, and which are therefore not subject to depletion through over-harvesting. The vast majority of marine species fall into this category. In American waters, there are 959 fish stocks, but only around 130 of these are considered commercially valuable, although almost half of the fish stocks have federal management plans (NOAA 2014). The other species are under an anarchic system.

Government policies encourage unsustainable harvests

In the US and worldwide, the anarchic system began to break down in the 1960s. Until then, commercial fishing was guided by profits and losses. To the degree that fish were plentiful and relatively easy to catch, fishing was profitable and fishing fleets grew. But when fish became temporarily scarce, the returns from fishing fell, profits sank and fishers either left the industry or cut back. Fishing fleets didn't overharvest fish stocks more than temporarily, since if they did so they lost money.

Up through the 1960s, this was basically the way fisheries operated. At that time, a number of government and private studies argued that the world's marine resources were underutilised. For instance, a 1969 report requested by Congress, while noting that the total annual world harvest from the oceans stood at 50 million tonnes, estimated that with the equipment then available production could be expanded to 150–200 million tonnes – three to four times the then present levels. However, it went on to argue that if technology and investment weren't holding back the industry, 'far greater quantities of useful, marketable products could be harvested to meet the increasingly urgent world demand for protein foods ... [making it] more realistic to expect total annual production of marine food products (exclusive of aquaculture) to grow to 400 to 500 million tonnes before expansion costs become excessive' (CMSER 1969).

To understand just how wildly optimistic – and badly mistaken – the government's report was, one should note that in 2012 the total harvest from both inland and ocean fisheries was just under 160 million tonnes. The harvest of wild caught fish has largely plateaued since 1990 in the range of 82–91 million tonnes (FAO 2014). This is four to five times lower than the government estimated could be safely harvested from the sea with new equipment.

Backed by these studies, the government concluded that ocean resources in US waters were underfished. As a result, the federal government began to subsidise fishing in ways that encouraged the type of overfishing that never would have occurred under anarchy.[1] The subsidies included:

---

1   For example, the Marine Resources and Engineering Development Act of 1966 33 USC §§1101–1108, 17 June 1966, as amended 1966, 1968–70 and 1986, or the Magnuson–Stevens Fisheries Conservation Act, 1976, 16 USC §§1801–1882, 13 April 1976, as amended 1978–80, 1982–84, 1986–90, 1992–94 and 1996.

- Below-market-rate loans for fishermen who bought bigger boats and state-of-the-art equipment.
- Tax breaks for investment in new equipment.
- Grants to fishing harbours to improve and expand the number of mooring spaces, and to purchase the latest equipment for fish warehouses.
- Grants and below-market-rate loans to fish processors for larger, newer plants.
- Tax credits for and/or waivers on taxes on marine fuel.

The result of these programmes was more fishing boats chasing fewer fish. In the late 1970s and early 1980s, there was a spectacular expansion in the US fishing fleet. Of all the fishing vessels built in the past 50 years, more than half were built during the decade from 1973 to 1984.

Worldwide, the story was much the same. Throughout the 1980s, while the number of fish declined, government subsidies caused the world's fishing fleet to more than triple. The Chinese fleet alone more than quadrupled between 1970 and 1990.[2]

Subsidies: what kind and how much?

Not all fishing subsidies are equal. Fishing subsidies are generally divided into three types: beneficial, capacity-enhancing and ambiguous. According to Per Oceana:

---

2 'Statistics for China's fishing output credible: official', *People's Daily*, 18 December 2001.

Beneficial subsidies enhance the growth of fish stocks through conservation, monitoring and control of catch rates. Beneficial subsidies include programs such as fisheries management, research and Marine Protected Areas.

Capacity-enhancing subsidies stimulate overcapacity and overfishing through artificially increased profits that further stimulate effort and compound resource overexploitation problems. These include programs such as fuel subsidies, boat construction and modernization, fishing port construction and renovation, price and marketing support, processing and storage infrastructure, fishery development projects, tax exemptions and foreign access agreements.

Ambiguous subsidies can lead to positive or negative impacts on the fishery resource depending on the design of the program. Some examples include fisher assistance programs, decommissioning and buyback programs and community development programs.[3]

---

3   The labels *beneficial*, *capacity-enhancing* (which seems to mean *harmful*) and *ambiguous* are common in the fisheries subsidies literature. Their use here should not be taken as an endorsement by the author. The labels are tied to the supposed ability of the subsidies to harm or enhance fish stocks, not an assessment of the local, regional or global economic effects of the subsidies. The author suspects that even the subsidies in the beneficial category actually are economically harmful and possibly even harmful to the fisheries themselves since, by intervening to enhance fish stocks, they help to keep inefficient, capacity-enhancing, subsidy-receiving commercial operators catching fish. Absent beneficial interventions, fish stocks would be lower (at least in the short term), leaving less fish to catch which might put the least efficient operators out of business, over time reducing fishing capacity. To be clear, the author supports an end to all fishing subsidies.

In most developed and developing countries, more money flows to capacity-enhancing subsidies than to the other two categories combined, with fuel subsidies leading the way at 22 per cent of all subsidies. Depending upon the country, fuel subsidies range from direct payments or credits to buy fuel, to waiving taxation on fuel bought by commercial fishing operators (Global Ocean Commission 2013).

Though fishing subsidies are growing in developing countries, developed countries still account for the vast majority of the world's fishing subsidies, some 65 per cent of the total.

Japan, China and the EU are the top three subsidisers of fisheries. Of the ten largest developed nations, only two, the US and Canada, spend more on 'beneficial' subsidies than capacity-enhancing ones. As of 2009, the US spent approximately $1.1 billion on 'beneficial' subsidies including fisheries management, research and development, compared to $342 million on capacity enhancement (Sea Around Us Project 2010).

Worldwide fishing subsidies amount to between $27 billion and $35 billion dollars a year (Global Ocean Commissions 2013). Carl Safina estimated that when fishing subsidies are combined with other wasteful actions in response to unwise regulations, the fishing industry was spending $124 billion a year to catch $70 billion worth of fish (Safina 1995). Whether one takes it as a good or bad sign, the annual economic losses of global fisheries have seemingly stabilised at approximately $50 billion since 1995 (Oceana 2011).

Misguided government responses to the fisheries decline

The fisheries would have been much better off if the government had done nothing over the past three decades. Left on its own, no private company or industry could continue to operate with such high losses.

In ancient times, Emperor Nero reportedly fiddled while Rome burned. The US and other governments, rather than fiddling as fish stocks were overharvested, actively poured gasoline on the fire by preventing a normal contraction of the industry.

The primary government response to the threat from overfishing was the 1976 Magnuson Fishery Conservation and Management Act, which brought American waters under government control. There were two main features of this legislation.

Firstly, the act created 200-mile economic zones in American coastal waters that are exclusively accessible by American fishermen. Secondly, it divided American waters into eight fishery regions, each of which was placed under the authority of a regional fish council. There are currently 39 separate fishery management plans in place around the country. These councils were given the task of formulating and implementing fishery law in their region. Each of the eight councils drafts management plans specifically tailored to the breeding seasons, migration routes and current stock levels of fish species in their regions. To do this, the councils use command-and-control regulations, which are aimed at preserving stocks by placing restrictions on four distinct areas:

- the size of fishing vessels and types of nets/traps;
- the length and timing of the fishing season;
- the areas that are open to fishing;
- the amount (usually specified in tonnage) of particular species that can be kept.

Theoretically, by controlling the means of catching fish, the councils can limit the access that fishers have to natural resources. The goal is to allow the fisheries to renew themselves, ensuring their availability for future generations. However, the political process suffers from two weaknesses: it is itself a commons, and it does nothing to change the incentives facing fishers.

There are dozens of federal and state agencies administering more than 140 different laws regulating the use of marine resources in US coastal waters. These efforts are failing or having only limited success. As the government is itself a commons, people who support bad policies bear only a small part of their costs. Most of the costs are borne by others. On the other hand, people who support good policies reap only a small portion of the benefits, reducing their incentive to act. As a result, the pursuit of political self-interest all too often results in environmental harm, as it has in the fisheries.

For every council, the federal government assigns conservationists and fish biologists with specialised knowledge of the fish in the region to advise the council members in setting policies. However, in most councils, the majority of council members are themselves fishermen, and therefore have strong financial incentives to promote their own interests. Of those members who are not fishermen, many are politicians, who are primarily concerned with meeting the needs of the fishermen in their districts in order to secure their votes in the next election. Neither the fishermen nor the politicians have very strong incentives to adhere to the recommendations of the conservationists or fish biologists. The result has been fishery management plans that are more concerned with short-term profits and votes than with conserving fish.

In Europe, the problem is similar, with Oceana reporting that in 2010 EU fisheries ministers ignored scientists' advice and set the catch for managed fish stocks in the Atlantic 20 per cent higher than recommended levels (Oceana 2011).

Regulations implementing the Magnuson Fisheries Act encourage commercial fishers to catch the maximum amount of fish they can because they have no ownership over fish until they are actually caught. As a result, there is a 'race to fish', as fishermen use every tactic and tool available to extract the largest catch. While conservation may seem nice in principle, any fisherman who decides to conserve the resource by limiting his catch will lose fish to those who are not public spirited and conservation minded. Simply put, if the conservation-minded fisherman doesn't get the fish, the profit-minded one will, and the first fisherman will not be a fisherman for long.

The problem with regulations is that fishermen have an economic incentive to avoid and evade them – through both legal and illegal means:

- Prevented from fishing on some days, they make a greater effort to fish on days when fishing is allowed.
- Forced to use smaller boats, they use more of them.
- Forced to use smaller nets, they use those nets more often.
- In response to limits on the number of fish they can bring back to harbour, they continue to overfish – throwing the smallest ones overboard, before their return; for example, in US waters, over a million tonnes of dead fish are thrown back into the ocean every year (Murray 2004).

The lesson is that commercial fishers, like everyone else, are rational profit maximisers. The horrific impact of this combination of anarchy (on the open seas) and politically driven regulation (in US coastal waters) was described earlier. Both in US waters and worldwide, the world's fisheries are in decline and fish stocks have plummeted. But the harm caused by these policies is not limited to the fish populations themselves. Indeed, commercial and recreational fishers, the US economy and society in general all suffer ill effects from these policies.

Current policies harm commercial fishing operators

The regulatory approach to fisheries management is costly and inefficient. It leads to overcapacity and overcapitalisation (too many fishermen and too many boats relative to the number of fish). Commercial fishermen are also hurt when regulations idle them for much of the year, leaving them underemployed or on social assistance. Newly enacted equipment restrictions encourage them to fish during dangerous weather conditions and to run up huge debts, often leading to bankruptcy. Indeed, they may be driven to over-harvest fish in the current season to service their debt, thus reducing harvests for future seasons.

Below-cost fishing is also a problem in the EU. In the EU, subsidies are equal to approximately half of the overall value of the total catch. As in the US, China, Japan, Russia and many other countries, the fleets of many EU countries only continue to operate because of government support. It is estimated that the fishing fleet is as much as three times larger than what sustainable limits would allow. Indeed, subsidies to the fishing sector exceed the value of the total fish catch in 13 EU countries. Four of these countries (Austria, the Czech Republic, Hungary and Slovakia) do not have fishing ports and therefore have no income from fish landings. However, these countries still receive fishing subsidies, mostly for aquaculture or inland fishing.

Finland's subsidies were three times larger than the value of the landed catch, and Germany's subsidies were 1.5 times the value of the catch. With subsidies worth six times the value of the country's catch, Poland has by far the largest discrepancy (Oceana 2011).

Acting to reverse fisheries decline

The problem with the management of marine resources is not bad people but faulty institutions. If flawed government policies that create incentives to over-harvest marine fish are eliminated, then commercial fishers and society in general will benefit. By contrast, combining subsidies with incentives to overfish will result in a bleak future for both fish stocks and the fishing industry.

When one finds oneself in a hole, the first step to getting out is to stop digging. In order to reverse the decline in commercially valuable fish stocks, and the poor economic performance of the industry, governments must first end the subsidies that encourage overinvestment in capital stock, boats, nets, other equipment, ports, harbours, and so on, as well as the other subsidies that allow below-cost fishing to continue, such as fuel subsidies and income support.

Ending subsidies may be easier said than done, however. Despite a general recognition by governments that many of the most valuable fish stocks are overexploited and at serious risk of collapse, governments don't treat fisheries in isolation from other domestic social and economic problems such as unemployment, welfare, taxes, etc., or from geopolitical concerns. Indeed, as the OECD noted in 2006, 'Historically, fisheries subsidies have been used as social policy tools to address concerns such as regional coastal development, community support and unemployment in fishing communities' (OECD 2006).

The US has been attempting to reduce overfishing since 1976 with the Magnuson–Stevens Fisheries Conservation Act, yet the solutions adopted since then have only exacerbated the problem as shown above.

Although, over the past decade, some US fisheries have recovered and spending on 'beneficial' subsidies is now greater than on capacity-enhancing subsidies, such success may prove temporary. As evidence, in December 2013, during the Ninth World Trade Organization (WTO) Ministerial Conference in Bali, Indonesia, US Ambassador Michael Punke joined ambassadors from the Friends of Fish group to announce a shared commitment to restrict harmful fisheries subsidies. Friends of Fish comprises Argentina, Australia, Chile, Colombia, Costa Rica, Ecuador, Iceland, New Zealand, Norway, Pakistan, Peru, the Philippines and the US. As a group, they pledged to refrain from introducing or expanding subsidies that contribute to overfishing or overcapacity (United States Department of State 2013). Note that they did not announce plans to reduce existing subsidies.

Less than a year later, however, the situation seemed to have changed, at least for the US. In November 2014, the National Marine Fisheries Service in the US Department of Commerce announced that it might expand state support for the American fishing industry, in particular by increasing the amount of long-term loans and grants available to expand capacity and purchase new fishing boats and equipment. In a written statement, the Port of Seattle, the authority managing international trade and travel for the coastal city, said 'new vessel construction and upgrades could generate between $7 and $14 billion in domestic economic activity and thousands of new jobs in Washington state'. The port argues that the fishing industry requires new, safer vessels and that private commercial markets do not correctly evaluate the risk associated with the capital needs of the fishing industry (Biores 2014). One agency commits to take fisheries in one direction, while another pulls it back to the future with expanded subsidies.

The international situation is no better, and arguably even worse. Internationally, fisheries management falls under a number of different treaties and governing bodies, each concerned with overfishing and threats to the ocean fisheries, but also each with different primary organisational objectives and with varying degrees of management and oversight authorities and responsibilities.

Bilateral agreements with links to fisheries management include: the EU Maritime Fisheries Fund, the Common Fisheries Policy (affecting other countries but limited to fisheries management among EU member states) and the Transatlantic Trade and Investment Partnership (under negotiation). Multilateral or international agreements or bodies that touch on fisheries policy include: the World Trade Organization, the Rio+20 declaration, the Johannesburg Plan of Implementation adopted by the World Summit on Sustainable Development, and the FAO.

However, none of these governing bodies has true policing authority that could override domestic agendas or plans; and no common fishing policy has been developed that carries enforceable penalties for violating agreements already made. Thus far, goals have been more aspirational than actual. As Thomas Hobbes put it, 'Covenant, without the sword, are but words and of no strength to secure a man at all'.

End subsidies and tax breaks

When the Magnuson Fishery Act was enacted, America's fisheries were thought to be an inexhaustible and underdeveloped resource. Commercial fishers were given tax breaks and subsidies to attract more of them to the industry and to encourage them to fish more often. The result was rapid growth in the nation's fishing fleets and a rapid decline in the nation's fish. Scientists now recognise America's fisheries were never as abundant in the 1960s and 1970s as people thought. Fisheries were not underdeveloped then and they certainly are not today. The case for subsidies is not supported by science or economics; thus they should not continue.

In the US, a first step to help fish stocks and fishing profits improve would be for the federal government to end subsidies for fishermen to purchase boats and other equipment. In addition, the government should end price supports to increase the market value of fish, stop providing rebates on fuel and equipment and stop giving money to ports for the construction of docks and other harbour facilities.

To avoid the sudden shock to the industry of removing these supports, the government could gradually phase them out over the course of several years, a 20 per cent annual reduction in support from present levels over five years, for example.

Removing subsidies will help to make fishing an economically healthier industry. It is important to bear in mind that a significant percentage of the current fishing community would not be in the business at all if it were not for the government, because they would not be able to make a profit on the dwindling fish stocks. For those inefficient fishing operations, the removal of subsidies will remove the incentives to continue building boats and hiring deck hands and will replace them with incentives to operate more efficiently or look for employment elsewhere. For already efficient fishermen, the removal of subsidies will help to remove their less efficient competitors and allow them to expand their operations.

Replace the current regulatory system with a system of property rights

Beyond ending subsidies, new (to fisheries) institutions are required to best utilise ocean fisheries, allowing species at risk to recover while maintaining continued harvests of commercially valuable stocks at sustainable levels. Developing some form of ownership in ocean areas and fish stocks, where it has been tried, has succeeded in improving both the profitability of commercial fisheries and the recovery of fish stocks. Governments should expand the use of property rights to more fisheries and fish stocks.

While removing subsidies is an important step in restoring fisheries, it will not achieve its full potential unless the larger regulatory system of which it is a part is also changed. Even if subsidies are removed, the fishermen that remain will still have incentives to 'race to fish' as long as they are competing for access to a resource that they cannot own. Establishing a system of overlapping systems of property rights in different marine resources should help fish species to flourish as it has done for cattle, sheep, chickens and hogs. There is not one uniform way of applying property rights to all marine resources. Four approaches have been tried around the world that would serve as a good starting point for managing many of the US's most important marine resources:

- Allowing ownership of shore land that is covered with water at high tide as a way of managing clams, mussels and oysters.
- Allowing ownership of parcels of the ocean floor, so that individuals can create artificial reefs.
- Allowing individuals to 'fence off' areas of the ocean as a way of managing migratory fish.
- Creating tradable rights – Individual Transferable Quotas – that entitle fishermen to a certain portion of the catch.

In each case, the goal would not be to apply an inflexible, prefabricated management technique to all types of resources, but rather to experiment with property rights and find techniques best suited to individual resources. For those types of fish that have been successfully privatised in other countries, the US should adopt similar methods for the same fish species in its own waters. For example, the US could implement an ITQ system for lobster fisheries along the Eastern seaboard like the ones used in Australia and New Zealand. For fish such as tuna, which move in large schools, the US could use fenced-off, Australian-style fish farms. For other types of fish that have not been privatised elsewhere, the US should experiment to find property rights systems that are suited to each species' habitat, migratory patterns, etc. In every case, the principles guiding US fisheries policy should be experimentation and innovation.

Encourage other countries to cut subsidies and adopt similar property-based fisheries policies

The problems facing marine resources are not limited to American fisheries. Instead, they are part of a global trend of stagnating or declining fish stocks, reflecting decades of regulatory mismanagement by governments around the world. Unlike herds of cattle or flocks of chickens, schools of fish move in and out of the jurisdictions of different countries. Their population levels in various fisheries are interconnected. This means that effectively managing fish with property rights in one locale will not always work unless other countries follow suit. Otherwise, some fish species will thrive in American waters, only to be overfished when they migrate into Canadian or Latin American fishing grounds.

While the US needs to act even if other countries do not, the US should encourage all countries with fishing interests and industries to abolish their own fisheries subsidies and adopt property rights–based systems. Taken together, the implementation of these two measures abroad will help to ensure that (a) de-subsidised American fishermen are not put at a disadvantage by having to compete against government-supported foreign competitors in international seafood markets and that (b) fish will have protectors and defenders wherever they may go. This can be accomplished through bilateral agreements with individual states or through multilateral international conventions. The prospects for international cooperation on this score are good.

In some places extending property rights to ocean fisheries is already working. Indeed, altogether, since the early 1980s, 17 countries have introduced property rights for managing their fisheries and in each case the condition of the fish stocks and the profits of the fishers have improved significantly:

- In Iceland (see Chapter 5), after decades of unsuccessful attempts to restrict fishing through quotas, the government introduced property rights in the country's herring fisheries. As a result, the number of vessels fishing for herring fell from 200 in 1980 to 30 by 1995. Subsequently, catches fell to sustainable levels, even as the value of catches rose dramatically (Gissurarson 2000).

- In Australia, property rights are used to manage 15 species of fish – including orange roughy, bluefin tuna and lobster. Prior to the privatisation, Australia's bluefin tuna fisheries were in near collapse due to an oversized fishing fleet. Since the introduction of property rights in 1989, annual catches have fallen by 60 per cent, the average income of fishermen has increased dramatically, and the nation's tuna fisheries have become the most profitable in the Pacific (Newby et al. 2004).
- In the US, property rights are used for four fish stocks: the Atlantic bluefin tuna fishery, the mid-Atlantic surf clam fishery, the Alaskan halibut and sablefish fishery, and the South Atlantic wreckfish fishery. All four of the federal fisheries that have been privatised now have smaller fishing fleets, higher incomes for fishermen, and larger, healthier fish stocks.
- By the early 1960s, English rivers were among the most heavily polluted in Europe, with almost no remaining salmon stocks. To reverse this trend, landowners along the banks began using their traditional 'riparian rights' to take legal action against polluters under the common law. This has brought salmon and trout populations to levels not seen since the early nineteenth century. To take one example, the River Derwent in Derbyshire was one of the most heavily polluted rivers in England, with almost no remaining fish stocks. Following the use of property rights to challenge its polluters, the river is now one of the cleanest in Europe (see Bate 2001).

- Following the British example, the Canadian province of New Brunswick introduced a property rights scheme in its inland fisheries in the 1980s. As a result, it now has the healthiest salmon stocks in Canada.

Evidence from the areas that have privatised their marine resources indicates that, with fishermen no longer facing the perverse incentive to deplete fish stocks, populations should rebound. Additionally, property rights holders will have incentives to reduce the catch of sexually immature fish to ensure future populations; to reduce by-catch since the harvest and disposal of non-commercial fish wastes time and resources and can detrimentally affect the ecosystem; and, where possible, to enhance the marine environment to increase fish stocks.

Under the property rights approach, resource users homestead, purchase or are assigned ownership rights to the resource to be harvested. The rationale behind property rights in wildlife is that, unlike regulation, they create incentives to conserve fish for long-term exploitation and profit. As owners, fishermen will reap the benefits of wise use and bear the costs of overuse. Resources that are owned by individuals have protectors and defenders because their owners have a self-interest in maximising the value of their own property.

Privatisation of marine resources has worked where it has been tried. Indeed, even as government-operated fisheries continue to decline, privately owned fisheries in the US and other countries have prospered.

References

Bate, R. (2001) *Saving Our Streams: The Role of the Anglers' Conservation Association in Protecting English and Welsh Rivers*. London: Institute of Economic Affairs.

Biores (2014) US authorities mull new fisheries subsidies. International Centre for Trade and Sustainable Development, 5 November (http://www.ictsd.org/bridges-news/biores/news/us -authorities-mull-new-fisheries-subsidies).

CMSER (1969) *Our Nation and the Sea. A Plan for National Action*. Report of the Commission on Marine Science, Engineering and Resources. Washington, DC: United States Government Printing Office.

Food and Agriculture Organization of the United Nations (FAO) (2014) *The State of World Fisheries and Aquaculture 2014* (SOFIA) (http://www.fao.org/3/a-i3720e/index.html).

Gissurarson, H. (2000) *Overfishing: The Icelandic Solution*. London: Institute of Economic Affairs.

Global Ocean Commission (2013) Elimination of harmful fisheries subsidies affecting the high seas. Policy Options Paper 6, November.

Greenpeace (1996) *Fishing in Troubled Waters – The Global Fisheries Crisis* (http://archive.greenpeace.org/comms/fish/amaze .html).

Hampton, S., Sibert, J., Kleiber, P., Maunder, M. and Harley, S. (2005) Fisheries: decline of Pacific tuna populations exaggerated? *Nature* 434: 1–2.

Hardin, G. (1968) The tragedy of the commons. *Science* 162: 1243–48.

Hutchings, J. and Reynolds, J. (2004) Marine fish population collapses: consequences for recovery and extinction risk. *Bioscience* 54(4): 297–309.

Murray, M. (2004) Nation's ocean ecosystem threatened by industry. *Ventura County Star*, 19 April.

Norcliffe, G. (1999) John Cabot's legacy in Newfoundland: resource depletion and the resource cycle. *Geography* 84(2): 97–109.

Newby, J., Gooday, P. and Elliston, L. (2004) *Structural Adjustment in Australian Fisheries*, ABARE eReport 04.17. Prepared for the Fisheries Resources Research Fund (http://www.abareconomics.com/publications_html/fisheries/fisheries_04/er04_structure_fish.pdf).

NOAA (2013) *Status of Stocks 2013, Annual Report to Congress on the Status of U.S. Fisheries – 2013.* Silver Spring, MD: US Department of Commerce, NOAA, National Marine Fisheries Service (http://www.nmfs.noaa.gov/sfa/fisheries_eco/status_of_fisheries/archive/2013/status_of_stocks_2013_web.pdf).

NOAA (2014) *Status of Stocks 2014, Annual Report to Congress on the Status of U.S. Fisheries – 2014.* Silver Spring, MD: US Department of Commerce, NOAA, National Marine Fisheries Service (http://www.nmfs.noaa.gov/sfa/fisheries_eco/status_of_fisheries/archive/2014/2014_status_of_stocks_final_web.pdf).

Oceana (2011) The European Union and fishing subsidies. September (http://oceana.org/reports/european-union-and-fishing-subsidies).

OECD (2006) *Financial Support to Fisheries: Implications for Sustainable Development.* Paris: OECD.

Safina, C. (1995) World's imperiled fish (global fish declines). *Scientific American*, pp. 46–53 (http://www.seaweb.org/resources/articles/writings/safina6.php).

Sea Around Us Project (2010) Fisheries subsidies in the U.S. (http://www.seaaroundus.org/Subsidy/default.aspx?GeoEntityID=221).

Shiffman, D. (2014) Predatory fish have declined by two thirds in the 20th century. *Scientific American*, 20 October (http://www.scientificamerican.com/article/predatory-fish-have-declined-by-two-thirds-in-the–20th-century/).

United States State Department (2013) U.S. supports end of harmful fisheries subsidies (http://iipdigital.usembassy.gov/st/english/inbrief/2013/12/20131208288614.html#axzz3RvxL3bDz).

Walters, C. (2005) Exploratory assessment of historical recruitment patterns using relative abundance and catch data. *Canadian Journal of Fisheries and Aquatic Sciences* 62: 1985–90.

3 THE EUROPEAN COMMON FISHERIES POLICY[1]

Rachel Tingle

The EU Common Fisheries Policy has not worked and is not working for fish, fishermen, the marine environment, coastal communities, or consumers. The system is broken and the 2012 reform process is our best, last chance to fix it.

NUTFA and Greenpeace, 2012[2]

As other contributors in this book have explained, the management of maritime fish stocks and fishing poses considerable problems for policy makers of any country because of the problem of the commons. In open access waters, this means that each fisherman will fish as intensively as possible because prudent fishing by one fisherman to protect

---

1 An earlier version of this paper appeared as 'Freedom for fisheries?' in Minford and Shackleton (2016).

2 Joint declaration between Greenpeace, NUTFA, UK Fishermen's Associations and fishermen on the reform of the EU Common Fisheries Policy. (NUTFA, the 'New Under Ten Fishermen's Association', is a UK campaigning organisation representing commercial fishermen with boats less than 10 metres in length and/or not belonging to the large producer organisations.) (http://www.greenpeace.org.uk/media/reports/manifesto -fair-fisheries).

the stock will almost certainly only lead to larger catches by other fishermen. This can result in overfishing: that is, fishing at a higher level than is sustainable biologically, referred to as the maximum sustainable yield (MSY), which leads to a depletion and possible destruction of the fish stocks on which the fishermen's livelihood depends.

Economic theory suggests that the best solution to the problem of the commons is to make it possible to exclude people from consuming the resource by assigning it with property rights, but in the case of sea fisheries this is not an easy matter. To start with there has to be an assignment of property rights over the seas, and then there has to be some way of assigning property rights (or at least 'harvesting rights')[3] over the fish swimming in these seas. (Other problems arise from the fact that fish stocks can migrate over national jurisdictions and so fisheries' management may require international cooperation and mutual recognition between nations of fishing rights awarded.)

These issues would have had to be faced by the UK government or the devolved administrations if the management of our maritime fisheries were in national hands. Ever since 1973, however, when the UK joined the European Economic Community (EEC), property rights over the fish in the seas around the UK have been ceded to the Community and, since then, almost all aspects of fisheries have been managed through the European Common Fisheries

---

3   Harvesting rights are the right to take so much of the resource over a certain period and are not normally assigned in perpetuity.

Policy (CFP) as laid down in a highly detailed body of EEC, and subsequently European Union (EU), law.[4]

For more than thirty years such law has included regulations supposedly designed to conserve fish in EU waters. Nevertheless, by 2008 the European Commission itself estimated that, of the stocks of fish for which information was available, 80 per cent were being fished above MSY, compared with a global average of 25 per cent. Worse still, 30 per cent of these EU stocks being fished beyond MSY were outside safe biological limits, meaning that stocks might be unable to recover (COM 2008).[5]

Alongside this, the contribution of the fishing industry (that is, fishing, fish processing and aquaculture) to EU GDP fell from 1 per cent in the early 1970s to less than half a percent in 2009, and the number of people engaged in the industry throughout the EU fell from 1.2 million in 1970 to about 400,000 in 2009 (El-Agraa 2011).[6] This chapter traces the evolution of the CFP since its earliest days to explain how this decline has come about and looks at whether the EU's latest reform of the CFP, which came into effect on 1 January 2014, is likely to improve matters. It also

---

4   This also relates to the relationship with non-EU countries regarding access to fish, external trade in fish products, and the development and management of aquaculture, none of which are covered in this chapter.

5   The comparative figures for overfishing in those countries with which the Commission considered the EU should be on a par with were 25 per cent for the US, 40 per cent for Australia and 15 per cent for New Zealand.

6   This decline is even sharper than it might seem, since the 1970 figures apply to the EU15 (the countries that made up the EU before the 2004 enlargement), whereas the 2009 figures apply to the EU27.

examines the implications for the fisheries industry of the UK's June 2016 decision to leave the EU.

The history of the CFP is a complex one, but it falls fairly clearly into six time periods, which will be discussed briefly in turn.

1957–69: The conception and early development of the CFP

The rather dubious legal origins of the CFP are derived from the 1957 Treaty of Rome. This stated that there should be a common market in agriculture accompanied by a common agricultural policy (the CAP) and, almost accidentally, defined agriculture to include the products of fisheries.[7] In the early years the EEC Commission focused its attention on developing common policies for agriculture and ignored fisheries, but the development, from 1962 onwards, of a common market in fish (which entailed the removal of EEC internal barriers to trade and the implementation of a common external tariff) had implications for the six individual member states. In particular France and Italy, which both had fairly inefficient fishing sectors previously protected by high import

---

7   It was not until the extensive amendment of the 1957 Treaty by the 1992 Treaty on the Functioning of the European Union (the 'Maastricht Treaty') when Article 3 was amended to read, 'the activities of the Community shall include ... a common policy in the sphere of agriculture *and fisheries*' that the legal basis for the CFP became completely unambiguous. For full details of the legal basis for the EEC/EU competences over fisheries and fish products, see Churchill and Owen (2010).

tariffs, were faced with steeply rising fish imports.[8] These threatened domestic producers' profitability. As a result, their governments began to agitate for a structural fund to provide aid to enable the modernisation of their fishing fleets, as well as a system of price support for fish products, similar to that of the CAP. In 1966 the Commission responded with quite detailed proposals for common policies on fisheries (COM 1966) but for some time the competing interests of member states prevented much being done.[9] This only changed in 1970 with the application to join the EEC of the UK and three other nations (Norway, Denmark and Ireland) which had either big fishing industries or significant coastlines.[10] In response, the six scrambled to establish an *acquis communautaire* (body of Community law) in the area of fishing which the new accession nations would have to accept if they were to join, and which amounted to nothing less than a resource grab.[11]

---

8 According to Wise (1984), fish imports into France rose from 95,000 tonnes in 1957 to 282,000 tonnes in 1966.

9 Germany, Belgium and the Netherlands, all with relatively small but efficient fishing industries, saw no reason to spend money making the fishing industries of France and Italy more competitive with themselves.

10 Ireland has a long coastline and thus had potential legal claims to sovereignty over a large area of sea, but at that time had a relatively small fishing industry.

11 It appears this move was initiated by the French and there were some concerns among the other EEC member states whether it was legal under the wording of the 1957 Treaty. For a full discussion of how the accession countries were 'ambushed' over access to fishing grounds, see Booker and North (2005: 180–92).

1970–82: The establishment of common Community waters

On 30 June 1970, just hours before the beginning of the formal accession negotiations, the EEC Council of Ministers hurriedly agreed two regulations which formed the basis of the first fully fledged CFP. Council Regulation 2142/70 established the common organisation of fisheries markets, encouraging fishermen to band together to form producers' organisations (POs) that would centralise market supply in major centres and oversee quality and marketing. It also set up a market intervention system with the aim of establishing price floors for fish similar to the price-support system of the CAP. The other Regulation (2141/70) met demands for structural aid for the industry by providing access to the European Agricultural Guidance and Guarantee Fund (EAGGF) for money to modernise fishing fleets.[12] Most significantly, however, it established the principle of *equal access* to fishing grounds, by stating (Council 1976):[13]

> Rules applied by each Member State in respect of fishing in the maritime waters coming under its sovereignty or within its jurisdiction shall not lead to differences in treatment of other Member States. Member States shall ensure in particular equal conditions of access to and

---

12 Under these proposals the equivalent of $15 million was made available over the next three to five years to help fishermen in France and Italy upgrade their boats.

13 The regulations were reissued in 1976 as Council Regulation (EEC) 101/76 in English; the original regulations were only in French.

use of the fishing grounds situated in the waters referred to in the preceding subparagraph for all fishing vessels flying the flag of a Member State and registered in Community territory.

This principle, which lies at the heart of the CFP, gave boats registered in one member state the same access to the maritime fishing grounds of any other member state as boats registered in that state; it meant that the member states would no longer have control over their own national fishing grounds, rather fishing waters would be a common Community resource, open to exploitation by all member states. By essentially *widening* property rights over the seas, this posed obvious dangers of increased overfishing, particularly as the initial CFP proposals contained no conservation measures for fish stocks.[14]

At the time Regulation 2141/70 was adopted, national sovereignty over fishing waters in Europe was largely governed by the 1964 European Fisheries Convention, which had given coastal states sovereignty over waters 12 nautical miles (nm) out to sea from their 'baselines'.[15] These 12 miles were divided into a 0–6 nm zone in which the coastal state had exclusive fishing rights, and a less exclusive 6–12 nm zone in which those foreign states

---

14 At the last minute a bland supplementary preamble was added to Reg. 2142/70 simply stating that 'implementation of the common organization must also take account of the fact that it is in the Community interest to preserve fishing grounds as far as possible'.

15 The low water mark on the shore, or in the case of bays, a straight line drawn across the bay.

which had 'habitually fished' in this zone between 1953 and 1962 could also fish in the same areas, and roughly at the same rate, as they had previously. Outside these territorial waters lay the high seas over which no nation had exclusive fishing rights. Initially, then, the EEC equal access principle legally applied only to the 0–12 nm zones. However, there was huge resistance to this proposal from the accession nations who were essentially being asked to give up control over an important and valuable national resource. Largely because of fears about the potential cost of this to their fishing industry, the Norwegian fisheries minister resigned in protest and the Norwegians decided in a subsequent referendum in September 1972 (and again in 1994) not to join the Community after all.[16]

The UK prime minister at the time, Edward Heath, tried for months to negotiate a change to the regulations, misleading the general public in the process that national sovereignty over fishing in UK territorial waters would not be conceded.[17] But, regarding the interests of a few thousand fishermen of less importance than gaining access to

16 It was also one of the main reasons why Greenland, having gained autonomy from Denmark, withdrew from the EEC in 1985. The CFP also proved a stumbling block to Iceland, which applied for EU membership in 2009, but withdrew the application in 2015.

17 In July 1971 a White Paper, *The United Kingdom and the European Communities* (Cmnd. 4715), was distributed to every household in the UK. It stated, incorrectly, 'The Government is determined to secure proper safeguards for the British fishing industry. The Community has recognised the need to change its fisheries policy for an enlarged Community of Ten, particularly in regard to access to fishing grounds.'

the EEC common market,[18] Heath eventually agreed to the compromise that the right to equal access of UK territorial waters should be partly 'derogated' (put off) for a transitional 10-year period until 1983 when it would be reviewed again. During this period equal access would not be allowed in the 0–6 nm zone, or in those parts of the 6–12 nm zone where it was deemed that coastal communities were especially dependent upon fishing.[19] It was this derogation which allowed the UK's principal negotiator for the Heath government in the EEC accession talks, Geoffrey Ripon MP, to claim, quite inaccurately, in the House of Commons on 13 December 1971 that 'we retain full jurisdiction of the whole of our coastal waters up to twelve miles. ... these are not just transitional arrangements which automatically lapse at the end of a fixed period' (Booker and North 1996: 192). In the 1983 reforms of the CFP this derogation was in fact extended back to the full 12 nm zone as a means of protecting coastal communities and, ironically, because of the recognition that there was better fish conservation in these waters, this derogation was renewed again in 2003 and, most recently, in the 2013 CFP reform.[20] Because of this, the UK still retains exclusive national fishing rights in the 0–6 nm zone and retains exclusive rights in *parts* of the

18 Christopher Booker and Richard North claim to have learnt from someone close to the negotiations that Heath's calculation in 1971 was that there were 'only 22,000 British fishermen' and they were not 'politically significant' (Booker and North 1996: 80).

19 Negotiations on that principle eventually excluded about one third of the British coastline from equal access, although the historic rights of other member states to fish in these areas remained as before (Wise 1984).

20 See Regulation (EU) No. 1830/2013, Preamble (19).

6–12 nm zone but this does not exclude these zones from other aspects of CFP regulation.[21]

In any case, the principle of equal access is of great significance beyond the 12 nm zone (a fact which prompted the Norwegian decision). By the 1970s some coastal nations had extended their property rights over marine resources up to 200 nm from their baseline[22] and, although this was not fully legalised until the 1982 UN Convention on the Law of the Sea, it was already clear by the mid 1970s that such 200-mile Exclusive Economic Zones (EEZs) would almost certainly be upheld in international law. Iceland established a 200 nm EEZ in 1975, followed by the US, Canada and Norway in 1977. This had profound consequences for northern European fisheries, especially the UK distant-water fleets, based in Scotland and northeast England, which had traditionally fished in these waters and which, from then on, would only be able to do so by negotiation and at reduced levels.[23] So, this trend towards

---

21 For instance, since conservation measures for 'marine biological resources' are an exclusive EU competence, member states can only implement fish conservation measures in inshore waters to the extent that EU rules allow them to do so. CFP regulations relating to fishing vessels also apply, as do regulations concerning recreational fishing (such as the limits imposed since 2015 on angling for sea bass around the UK coast). In addition, to CFP regulations, inshore waters are also subject to separate EU environmental legislation.

22 Or, where the coastlines of two nations are closer than 400 nm, to the median point between them.

23 The negotiating text drafted at the conclusion of the third session of the United Nations Conference on the Law of the Sea (UNCLOS) in 1975 stated that a coastal state would only be obliged to grant other states access to exploit the proportion of the available fish catch it was unable or unwilling to catch itself.

200 mile EEZs meant a significant diversion of fishing effort, not only by Community fishing fleets, but also by similarly affected third-party states, into the northern waters around the EEC.

In 1976, responding to this perceived double threat on fish stocks, the EEC agreed that member states with coastlines bordering the North Sea and the North Atlantic should themselves simultaneously adopt 200 nm fishing zones on 1 January 1977. This was done by national legislation in each member state: in the case of the UK by the Fishing Limits Act 1976.[24] Because of the equal access provision, however, as Figure 3 shows this essentially extended EEC property rights over a vast area of sea. (The dark and mid grey areas are part of the EU's 'common pool', of which the mid grey area is the UK's national EEZ. The light grey areas to the north of the UK are the EEZs belonging to Norway, the Faroe Isles and Iceland.)

Since by that time it was becoming obvious that many European fish stocks were overfished, two crucial questions immediately presented themselves: firstly, how to limit catches in order that stocks might be conserved and, secondly, how to allocate these limited fishing opportunities between the member states. A related third issue was how to shrink the capacity of the Community fishing fleet (both in terms of tonnage and engine power), which was now recognised as being too large in relation

---

24 In fact, the UK would have created a 200 mile fishing zone unilaterally if need be: see *Hansard*, 20 October 1976, col. 1459. Fishing limits were also later extended in the West Atlantic, the Skagerrak and Kattegat and the Baltic, but not in the Mediterranean.

Figure 3   Exclusive Economic Zones around the British Isles

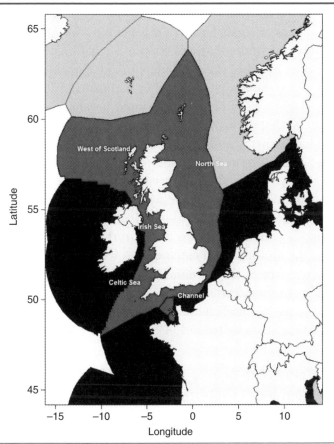

to the fishing opportunities – a problem made worse by the provision of European structural funds to modernise the fleet.

In 1976, the Commission came up with detailed proposals to address these issues (COM 1976). Given how much was at stake, however, it is perhaps not surprising that it took more than six years of squabbling between the nine member states to come to any agreement. Britain and Ireland argued that they had contributed by far the largest areas of sea and should be able to reserve some of that exclusively for their own fishermen (the UK fishing industry and the Irish government argued that these exclusive national fishing zones should be 50 nm or more), whereas the other seven member states insisted that equal access should apply throughout the new 200 mile zones. Britain and Ireland lost the argument over equal access but went on to claim that they should at least be allocated fish catch quotas which would reflect their shares of the 'community pond'. This was also rejected. Eventually the EEC crawled its way to a reformed and more comprehensive (but not necessarily effective) CFP that was adopted in 1983.

1983–92: The development of a fisheries management system

The 'basic' regulation of the 1983 CFP (Council 1983a) defined the objectives of the new system as being to 'ensure the protection of fishing grounds, the conservation of the biological resources of the sea and their balanced exploitation on a lasting basis and in appropriate economic and social conditions'. The main means for attempting to do this would be via the setting of an annual

Total Allowable Catch (TAC) for each of the main com-mercial fish stocks.[25] This was to be formulated initially by the Commission in the light of available scientific advice,[26] and agreed by the Council of (fishery) Ministers. These TACs would then be divided into *national* quotas for each fish stock. The regulation also gave the EEC the legal powers to introduce other 'technical' conservation measures which included such things as closing areas of the sea to fishing at certain times of the year to protect spawning and immature fish; restrictions on the use of fishing gear, such as type of nets used; and the minimum size of fish which could be landed.

To ensure all this was implemented, the CFP introduced control measures to police the system: these required all EEC skippers of boats over 10 metres to maintain stand-ardised log-books in which to record details of their catch; all member states to establish an inspectorate to check on fish landings; and set up a small multinational team of fisheries inspectors (originally 13, now 25) within the Commission to do spot checks on the national procedures.

Clearly, so far as national interests were concerned the most important aspect of the policy was how the division of

---

25 That is, fish species in certain defined areas of the sea – thus, sole in dif-ferent areas of the North Sea, for example, are regarded as a different fish stock from that off the West Coast of Scotland.

26 The basic regulation provided for the establishment of a Scientific and Technical Committee for Fisheries, now the Scientific, Technical and Economic Committee for Fisheries (STECF), to provide this information. The Commission also makes use of scientific information provided by the International Council for the Exploration of the Sea (ICES).

TACs into national quotas would be made.[27] The 1983 'basic regulation' 170/83 stated this should be on the basis of 'relative stability', which meant that the proportional share of the catch of each fish stock taken by any EEC member state should stay roughly the same. After intense negotiations it was decided this would be based on the average of past catches in the reference period 1973–78, with some adjustment under the so-called 'Hague Preferences' to give preferential treatment for regions particularly dependent upon fishing (some northern parts of the UK,[28] Greenland and Ireland) and reflect the loss of catches by distant-water fleets as a result of the introduction of the 200 mile fishing zones by Norway and Iceland. Because of this, the relative stability principle has had a huge part to play ever since in determining the fortunes of national fishing industries. In the case of the UK, because so much British fishing during the reference period had been in distant waters (particularly waters now within the Icelandic EEZ), it ended up with quota of just 37 per cent of the Community total by weight and, because this was skewed heavily towards lower value fish, only 13 per cent in cash terms (Booker and North 2005: 151).[29]

---

27 The very important related issue of how to divide national quotas among its fishermen was left for each member state to decide and, over the years, quite different methods have emerged. Some of these (principally Denmark and the Netherlands) now take a market-orientated 'rights-based' approach (see COM 2009b). For the present UK method, see the appendix to this chapter.

28 Northern Ireland, the Isle of Man, Scotland and the northeast coast of England between Bridlington and Berwick.

29 Similarly Ireland, which in the 1970s had an underdeveloped industry fishing almost entirely in inshore water, ended up with quota amounting to a mere 4.4 per cent of the total.

In spite of these measures, however, it was clear by the beginning of the 1990s that the CFP was failing in its management of fish stocks. Four main problems (of many) can be identified, all largely the product of the tragedy of the commons playing itself out in new ways. In spite of a series of reforms, these problems have been a feature of the CFP ever since. The first was the fact that effective implementation depended on fishermen's compliance with technical conservation measures and their keeping of accurate details of fish catches and landings,[30] as well as determined monitoring and policing of the system by member states, including halting the catch of particular fish stocks once national quota limits had been reached. Since it was in the economic interest of both fishermen and member states not to comply, many did not, particularly as at that time virtually no penalties were imposed on member states breaching their quota allocation or failing to comply with technical conservation measures. As the Commission admitted in 1991, 'compliance with TACs and quotas has been very limited' and 'the effectiveness of ... the penalties applied to the member states (is) virtually nil' (COM 1991: 22–23).

A second problem was that the TACs were set at too high a level. There were two reasons for this. Firstly, in advising on TACs the Commission lacked accurate data on fish catches (and therefore fish stocks) and also had inadequate scientific advice. Secondly, in the Council meetings, fishery ministers regularly pushed TACs to levels above

---

30 Including in non-EEC ports, and offloading at sea into other vessels.

those advised by the Commission in order to avoid their own national quotas from being cut. In effect, it was now the fishery ministers who were causing the problem of the commons, rather than the fishermen themselves.

The third major problem was the fact that TAC limits attempt to control fish *landings* not the number of fish *caught*, including those discarded (usually dead) back into the sea. This practice of discarding arises in many situations, including when juvenile fish are caught under the specified legal landing size; when legal but smallish fish are discarded in favour of higher-value larger fish, a practice known as 'high-grading'; and, in mixed fisheries, when species of fish are caught as a 'by-catch' to the main target fish and are either considered uneconomic to land or there is no available quota for them. The Commission was aware at that time that hundreds of thousands of tonnes of fish were being discarded but regarded it as an 'accepted evil' (COM 1991: 19–20).

The fourth problem was that, in spite of an awareness that the size of the Community fishing fleet needed to correspond to fishing opportunities, the EAGGF was still providing funds for 'economically appropriate expansion' and modernisation of the fishing industry, which subsidised 25–50 per cent of the costs of such investment (Council 1983b). From 1987 onwards, targets in the form of Multiannual Guidance Programmes (MAGPs) were introduced to reduce fleet tonnage and engine power (Council 1986, 1990). However, as with the TACs, the Council set these at levels above those advised by the Commission. The result of this was that, over the period 1983–91, fishing capacity actually *increased,* providing a strong economic incentive to continue to fish above quota.

All of these problems were exacerbated by the entry of Portugal and, more especially, Spain into the EEC in 1986. At that time Spanish fishermen had a fleet approximately three-quarters of the size by tonnage of the total of all the other EEC members. However, they added little to Community fish stocks, as their destructive fishing methods had virtually exhausted their own waters. To avoid Spain's complete disruption of the CFP, complex transitional arrangements were put in place, under which only a limited number of Spanish vessels would be allowed access to Community fishing grounds, and then not before 1 January 1995. It was planned that full integration would only take place in 2003.

In return for this delay Spain was given substantial aid from the EAGGF supposedly to reduce the size of the fleet, but much of which was actually used to modernise boats and hence increase their fishing capacity. Spanish fishermen also circumvented the interim ban from wider Community waters by using EAGGF and national subsidies to buy fishing businesses in other EEC countries, particularly the UK and France, and so qualify for a share of those countries' quota – a legal practice under the free movement of capital and freedom of establishment Community rules, even though the boats might be manned by Spanish fishermen and fish caught landed in Spain (Lequesne 2000). The UK attempted to stamp on this practice of 'quota hopping' through the Merchant Shipping Act 1988, which imposed nationality requirements on vessels seeking to benefit from

UK fishing quota. However, in a series of legal cases involving the Spanish-owned Factortame Ltd fishing company, the UK action was ruled by the European Court of Justice to be in breach of Community law.[31] In any case, faced with the Spanish threat to veto the 1995 EU enlargement (when Sweden, Finland and Austria joined), from 1996 the Spanish fleet was allowed equal access to EU waters, so putting further pressure on fish stocks. The EU has also spent a great of money deal since then, buying fishing opportunities in non-EU waters to accommodate the Spanish fishing fleet.

1993–2002: The introduction of vessel licensing and effort controls

In December 1992 the CFP was changed again with the adoption of a new 'basic' Regulation 3760/92 (Council 1992), replacing Regulation 170/83. The reforms included:

- multi-annual plans (MAPs) for fisheries management, in the hope that these would avoid dramatic variations in TACs and so allow the industry to plan ahead better;
- a multi-species approach to the setting of TACs to take more account of the impact of fishing on other fish stocks in mixed fisheries;

---

31 In spite of the requirement, introduced in 1999, that British registered fishing vessels over 10 metres in length and landing over two tonnes of quota stocks annually must demonstrate an economic link with fishing communities in the UK, a number of vessels fishing against UK quota continue to be part or wholly owned by non-UK citizens (mainly Spanish or Dutch).

- mandatory licensing of all Community fishing vessels; and regulation of fishing 'effort'[32] instead of, or in addition to, the TAC limits.

None of this did much to improve fish conservation or the economic health of the fisheries sector. As the Commission's Green Paper (COM 2001) noted, there had been limited progress in adopting multi-annual approaches. Effort management had proved unsuccessful, largely because it too was subject to bargaining within the Council,[33] who continued to systematically fix both TACs and MAGPs above levels proposed by the Commission. In addition there remained considerable variations between member states in the enforcement of the system and the imposition of penalties for infringement. Excess fleet capacity was a significant problem, particularly as structural aid, provided under the Financial Instrument for Fisheries Guidance (FIFG)[34] continued to enable fleet modernisation: this, because of technological creep through improved fishing gear, was increasing the ability to harvest fish more than just fleet tonnage and engine power might suggest. As a result, many fish stocks,

32 That is, the product of the capacity of a fishing vessel and its activity, normally expressed in terms of days allowed at sea.

33 One of the constant criticisms made by the industry about effort management is that it has introduced yet more complex regulatory micro-management into the system and yet, because of the numerous derogations negotiated in the Council of the EU, it has so far proved to be a very ineffective conservation measure.

34 This had taken over from the EAGGF as the structural fund for the CFP in 1994.

particularly demersal species such as cod, hake and whiting, were on average 90 per cent lower in the late 1990s than they had been in the early 1970s, and were now outside safe biological limits (COM 2002: 29). At the same time, much of the fisheries sector was characterised by poor profitability and steadily declining employment, with jobs in fish catching, for instance, declining by 22 per cent overall in the period 1990–98 (COM 2002: 3).

2003–13: Reform of the CFP

As the 2001 Green Paper shows, the staff at the Commission recognised the problems in the workings of the CFP (many of which continued to stem from the competing interests between EU member states and the inability of some member states to take the need for conservation measures seriously), but they have been fairly helpless to do anything about them (COM 2001). The Commission held extensive consultations with stakeholders in the industry over the period 1998–2002 and, in response to their deep dissatisfaction with the system, the Council adopted yet another new basic CFP Regulation, which came into force at the beginning of 2003 (Council 2002). The main aspects of this were the following:

- The adoption of multi-annual management or *recovery* plans for selected fish stocks (the latter, involving stocks deemed to be outside safe biological limits, might involve closing zones of sea to fishing for periods of time).

- The replacement of MAGPs with an 'entry/exit' regime whereby any new fishing capacity created with or without the use of EU public money should be matched with the withdrawal of at least the same amount of capacity.
- The introduction of tighter measures of control and enforcement. This included the installation of satellite-based monitoring systems on board all larger fishing vessels;[35] that fish could only be sold from a fishing vessel to registered buyers or at registered auctions (to help stamp out demand for 'black' or illegal non-quota fish); and tougher sanctions against infringements of the CFP to be applied both by member states against fishermen and by the EU Commission against member states.[36] It also allowed for a greater degree of cooperation between member states on enforcement matters, which led to the creation of a Community

---

35 The requirement applied to vessels longer than 18 metres from January 2004, and to vessels longer than 15 metres from January 2005 (Council 2002: Article 22).

36 The most notable example to date of sanctions by the Commission against a member state was in 2005 when the European Court of Justice imposed a €20 million fine on France for systematic capture and landing of under-sized hake, along with a penalty of €57 million for every six months until this was remedied. Some of the biggest fines imposed by the UK government against fishermen also involved hake, at that time regarded as on the verge of collapse. This was in July 2012 when a group of Spanish fishermen, fishing against UK quota, were found guilty of systematically failing to register the transfer of fish between vessels at sea and false readings given for weighing fish at sea (thereby underestimating how much they had fished against quota) over a period of 18 months in 2009 and 2010. They were fined £1.62 million by a court in Truro. See *The Guardian,* 26 July 2012.

Fisheries and Control Agency (CFCA), operational since 2007.

- The establishment of a Community Fleet Register (CFR), which means the Commission now holds regularly updated details on all commercial fishing boats, each of which is assigned a unique CFR number, so aiding both control and enforcement and the entry/exit regime.
- The establishment of Regional Advisory Councils (RACs) to feed stakeholder advice to the Commission. These would cover distinct fishing zones and be made up primarily of representatives of the fisheries sector, but they would also include other interested parties, such as environmental groups.

By 2008 six RACs had been set up.[37] They have generally been considered a success, enabling much greater input by those with detailed knowledge of local fishing conditions into the distant Brussels-based policy-making process. Other aspects of reform, however, failed. As far as the policy of multi-annual management of fisheries was concerned, by 2008 only four recovery plans and four management plans had been adopted, and annual TACs (by this time set for around 130 commercial fish stocks) continued to be the main instrument of fisheries management. These were still being set on average about 48 per cent higher than MSY (COM 2008: 331). Another significant problem was that,

---

37 These covered the North Sea, Pelagic fisheries, North-Western Waters, the Baltic Sea, South-Western Waters and the Long-Distance Fleet. An RAC covering the Mediterranean was established later.

even when scientific evidence pointed to the need for a big change in the TAC, existing EU rules meant they could not be varied by more than 15 per cent per annum.[38]

Crucially, too, member states lacked the political will to speed up a reduction in fishing capacity, particularly as the basic regulation had not set an overall reduction target, only that member states should 'achieve a stable and enduring balance between ... fishing capacity and their fishing opportunities' (Council 2002: Article 11). After the 2003 reforms, capacity continued to fall at roughly the same annual rate of between 2 and 3 per cent that it had over the previous decade. Even this small reduction was broadly offset by technological progress in fishing efficiency – some estimates put fishing overcapacity throughout the EU in 2008 at 40–50 per cent (House of Lords 2008: 23, 28).

This problem of overcapacity was exacerbated by the continued misuse of the EU structural fund, supposedly mainly intended to aid vessel decommissioning or alternative employment for fishing communities. Of the €3.2

---

38 Many in the industry argue that, because the state of fish stocks can change rapidly in response to water temperature or availability of food, TACs need to change rapidly too. As John Ashworth of the 'Restore Britain's Fish' campaign has argued (Ashworth 2016: 18), 'Now, just as cod moved off the Grand Banks because the water got too cold several years ago, so the cod are moving north in the North and Irish seas because the water is getting warmer. In their place, hake are moving in, for which we have very little quota – less than 12,000 tonnes in 2016 ... When you have a rigid system like the CFP, you might go several years in your area catching species for which you have a quota and then, suddenly, they disappear and in comes a species for which you have no or little quota. What do you do? Answer: you have to cheat to survive.' He and others have also pointed out the need to be able to close fisheries within a matter of hours when breeding is taking place.

billion provided by the FIFG between 2000 and 2006, approximately €1.5 billion went to Spain (three and a half times the total sum given to the UK, Germany and Poland combined). Spain used 60 per cent of this for vessel construction and modernisation, thereby further increasing the size and power of the Spanish fleet (Poseiden Aquatic Resource Management 2010).[39] Finally, a damning Special Report by the European Court of Auditors in 2007 found that the system of control, inspection and sanctions remained inadequate: catch data were neither complete nor reliable, the inspection system remained poor and few infringements were followed up with penalties sufficient to act as a deterrent. The report found the failure of the system was greatest in Spain, where, for example, quota monitoring ignored catches by vessels under 10 metres in length, even though such vessels accounted for 67 per cent of the fleet. As the European Union Committee of the UK's House of Lords concluded in its extensive 2008 report (House of Lords 2008: 6):

---

39 The FIFG was replaced in 2007 by the European Fisheries Fund (EFF), which provided financial assistance to the European fisheries sector of €4.3 billion over the period 2007–13, €1.12 billion of which went to Spain, compared with €134 million to the UK (COM 2014b). The EFF has now been replaced by the European Maritime and Fisheries Fund (EMFF), which, over the period 2014–20, will provide €5.7 billion to member states (total budget €6.4 billion). Over one fifth of this will go to Spain, compared with just 4.3 per cent to the UK. For details see http://ec.europa.eu/fisheries/cfp/emff/index_en.htm (accessed 8 July 2016). This problem of EFF funding potentially being used to increase the fishing capacity was also highlighted by a highly critical special report of CFP policies to reduce fleet capacity published by the European Court of Auditors in 2011 (European Court of Auditors 2011).

[O]n most indicators the 2002 reform of the Common Fisheries Policy has failed: overcapacity in the fishing fleets of the Member States, poor compliance, uneven enforcement, and a stiflingly prescriptive legislative process all persist, while fish stocks remain depleted.

In more recent years, however, there have been some signs of improvement in stock conservation. By 2009 about 41 per cent of pelagic fish and 29 per cent of demersal fish were being managed under long-term management plans, and their greater flexibility now enabled annual TACs to be varied by up to 30 per cent. TACs were also being set slightly closer to the scientific advice, although many were still above MSY. New monitoring and control procedures had been put in place, including better data collection and wider implementation of electronic logbooks, enabling real-time catch recordings (Council 2009). By 2010 it appeared that some fish stocks in the EU's northern waters were recovering. In 2014 the Commission reckoned that the percentage of stocks overfished in the North East Atlantic and adjacent waters had fallen to 41 per cent from 94 per cent in 2005, and the percentage of fish outside safe biological limits had fallen to 17 per cent from 26 per cent (COM 2014a). It is notable, however, that 93 per cent of the known fish stocks in the Mediterranean remain overfished (COM 2015).

In 2009 the Commission published yet another Green Paper inviting further debate on the ways the CFP might be more substantially reformed, stating 'this must not be yet another piecemeal, incremental reform but a sea change cutting to the core reasons behind the vicious circle in which Europe's fisheries have been trapped in recent decades' (COM 2009a: 7).[40] One of the most notable aspects of this was the Commission's recognition of the very poor economic health of much of the EU fisheries sector (in several member states the cost of fishing to the public budget in terms of national and EU aid actually exceeded the total value of the fish caught) and, in an attempt to improve this, its desire to see fishing opportunities set at levels which could maintain or restore stocks to MSY (COM 2009a: 7).[41]

The other urgent and related[42] matter was to reduce discards. There are hugely varying estimates of how bad discarding under the CFP has been, but a paper produced by the Commission in 2007 estimated that, for the period 2003–5, discard rates were running at 20–60 per cent of the catch weight for typical fisheries exploiting demersal fish and that, between 1990 and 2000, around 500,000–880,000 tonnes of fish were discarded annually just in the North Sea (COM 2007).[43] An estimate by NUFTA and Greenpeace (2008) suggested that around 1.3 million tonnes of fish were

40 For a detailed UK parliamentary discussion of the proposed reforms, see House of Commons (2010–12).

41 The other reason for this policy was that, at the 2002 World Summit on Sustainable Development, the EU had pledged to set fishing opportunities within MSY by 2015.

42 If stocks are fished beyond MSY, there are more likely to be fewer large mature fish so more discards through 'high-grading' may take place.

43 The EU's Scientific, Technical and Economic Committee for Fisheries (STECF) has systematically been collecting data under the data collection Regulation 1543/2000 (now the more stringent Regulation 199/2008) since 2002, and the 2003–5 discard rate is based on these figures; however, they did not then have data for all sea areas; earlier figures are from the UN's Food and Agriculture Organisation (FAO).

being discarded annually in the North East Atlantic, while a 2011 study reckoned that the value of the cod discarded in the North Sea, Eastern Channel and Skagerrak between 1963 and 2008 amounted to £2.7 billion (New Economics Foundation 2011). The Commission itself was keen to see an end to discards and by 2011 there was mounting public pressure, particularly in the UK, for an immediate end to the practice.[44] This demand was eventually supported by Maria Damanaki, the EU Commissioner for Maritime Affairs and Fish, but it was opposed in the June 2012 Council meeting by a number of fisheries ministers, including the French and Spanish.

2014 Onwards: last chance for the CFP?

Eventually a compromise on discards was reached and enshrined in the December 2013 CFP new basic Regulation (COM and Parliament 1380/2013), which came into force at the beginning of 2014. The key aspects of this reform were:

- From 2015 onwards, starting with pelagic fish, a ban on discards has gradually been introduced on a fishery-by-fishery basis. (A ban on the discarding of demersal fish in the North Sea and North East Atlantic came into effect at the beginning of 2016.) This is referred to as the 'landing obligation' and means that by 2019 virtually all fish subject to quota must

---

44 Spearheaded by the celebrity chef Hugh Fearnley-Whittingstall and his 'Fish-Fight' campaign, which is estimated to have attracted 700,000 supporters.

be logged and landed and will count against quota.[45] Small fish below 'minimum conservation reference size' (MCRS)[46] must be landed, and although they can be sold, this cannot be for human consumption. Because of the landing obligation, TACs are being raised slightly and the ability of member states to 'bank and borrow' against subsequent years' quota is increased from 5 to 10 per cent, and fishermen are being given some greater flexibility to buy or swap quota. There will also be money available from the EMFF, the new fisheries structural fund, to help facilitate the discard ban by enabling vessels to install new gear to reduce by-catches, and for POs to fund marketing campaigns to promote the consumption of lesser-known fish caught as by-catch.

- A legal commitment that maximum sustainable yield exploitation rates should be achieved by 2015 where possible, and at 2020 at the latest, for all fish stocks.

- A renewed commitment to the management of fish stocks under multi-annual plans which will be based on MSY targets and include conservation measures where necessary.

- A proposed new form of regional government, whereby member states that share fisheries at sea basin

---

45 There are some exceptions to the total ban on discards: discarding is still allowed for fish damaged by predators, some non-quota species, fish which might survive discarding, and a very low level of discarding where the cost of landing the fish would be disproportionately expensive, referred to as the 'de minimis' exception.

46 Essentially the new name for minimum landing size.

level shall, in consultation with the RACs (renamed Advisory Councils), make joint recommendations to the Commission. The role of the Advisory Councils is also strengthened and four new ones will be established.[47]

- Member States will be required to produce and publish an annual report on the capacity of their fleet, including whether there is any structural overcapacity. If there is, they will be required to produce an action plan with a clear timetable setting out how this will be addressed.

These changes, many of which came about as a result of pressure from the British, are significant. The ban on discards and the requirement to fish within MSY have been widely welcomed within the industry, although it is too soon to tell yet how strenuously these proposals will be implemented and enforced, and whether they have come in time. The World Wide Fund for Nature (WWF) has argued, for instance, that the delay until 2020 in fully implementing fishing at MSY may be too late to save some fish stocks (WWF 2013). The situation is particularly dire in the Mediterranean and Black Seas. Also, there is little doubt that, in the short term, the landing obligation will impose additional costs on fishermen as they will have to store fish onboard for which there will be little or no financial gain and, in cases where there are no markets at present for fish

---

47 These will cover markets, aquaculture, the Black Sea and the outermost regions of the EU.

below MCRS and for some by-catch, it may impose land-fill costs on public authorities.[48] Hopefully, in the longer term, the problem of catching fish below MCRS might be solved by greater use of selective gear. And the problems of by-catch might be alleviated by the use of a real-time quota exchange, as is used in Denmark, which means that (thanks to the mediation of private bodies called 'Fishpools') quotas can be swapped or bought on-line while fishermen are returning to harbour (Fresh Start 2013: 98).

Other problems remain. Although the regionalisation plan has been welcomed widely in the industry, many in the UK think this could go further. At present it is not truly an example of subsidiarity, delegating decision-making powers down to the Member States and relevant stakeholders. Indeed, it maintains, and may even increase, the involvement of the Brussels bureaucracy (HM Government 2014: 44–46; House of Commons 2010–12: 9–13). In addition, obligations under the Lisbon Treaty, which came into force in 2013, mean that CFP legislation (though not the settting of TACs) now has to be agreed by both the Council of the EU *and* the European Parliament. Legislative procedures surrounding the CFP may therefore be even more cumbersome than they have been to date and even more vulnerable to competing national interests.[49] Indeed, it is not obvious that the central

---

48 The Scottish Government, for instance, has acknowledged that some land-fill may take place. See http://www.gov.scot/Publications/2016/03/2058/6.

49 A row regarding which aspects of the CFP can be decided by the Council alone, and which now also have to be put before the European Parliament, has had to be ruled upon by the ECJ, so considerably delaying the drafting of new multi-annual plans for fisheries (HM Government 2014: 52).

Table 1   Landings of fish into the UK by UK and foreign vessels, 1938–2014 (thousand tonnes)

	1938	1948	1960	1970	1980	1990	2000	2010	2014
Demersal	808	924	759	779	484	337	246	149	151
Pelagic	295	288	128	204	319	268	152	230	229
Shellfish	32	29	28	56	70	98	128	141	144
Total	1,135	1,240	915	1,039	874	702	526	520	523

Source: UK Sea Fisheries Statistics 2014; Marine Management Organisation 2015.

problems of the CFP – the infighting to secure the highest possible TAC and hence quotas for individual member states, the varied levels of enforcement, the slow and bureaucratic micro-management from Brussels, and the misuse of the fishing structural funds – have been overcome.

European fisheries obviously need a policy which can deliver biologically and economically sustainable fishing over the longer term. It is widely agreed, however, that so far the CFP has signally failed to do this. During the forty years it has been subject to the CFP, the UK fishing industry has been in almost continuous decline as stocks have fallen and fishing opportunities have been restricted. As Table 1 shows, while landings by weight into UK ports of pelagic fish have fluctuated considerably since 1970 showing a slight overall increase of 12 per cent to 2014, those for the more valuable demersal fish have been in almost continuous decline, plummeting by 80.6 per cent from 778,600 tonnes in 1970 to 150,600 tonnes in 2014.[50]

---

50 These figures include landings into UK ports by non-UK-owned vessels, and exclude landings by UK-owned vessels into non-UK ports. For more details, see Marine Management Organisation (2015).

The size of the fleet has also fallen, both in response to the economic pressures resulting from falling catches and also because of EU and UK government encouragement to decommission vessels. Numbers of vessels fell from 8,667 in 1996 to 6,383 in 2014, with a resulting reduction in terms of capacity from 274,532 gross tonnage (GT) to 195,121 GT. The number of regular and part-time fishermen has shrunk too, from 19,044 in 1996 to 11,845 in 2014 and by nearly a half since 1970 (Marine Management Organisation 2015: Tables 2.1 and 2.6).[51] Neither the size of the fleet, nor the number of fishermen, so far indicates that they have stabilised. UK fish consumption per capita is falling, but, in spite of this, the industry is unable to satisfy demand. In 2014, the UK was a net importer of 221,000 tonnes of fish, with a value of £1.3 billion, equal to just under one third of total UK consumer expenditure on fish (ibid.: Tables 4.1 and 4.5).

It is difficult to avoid the conclusion that the UK would have done better to retain national control over its fisheries, as, for example, Norway, Greenland and Iceland have done. When the British House of Lords conducted its extensive and very critical review of the CFP in 2008 it dismissed the UK's withdrawal as a credible policy option by stating 'unilateral withdrawal would be incompatible with membership of the EU, while negotiated withdrawal would require unanimous agreement to a treaty amendment by all Member States' (House of Lords 2008). With the UK's decision to leave the EU, however, all

---

51 Part-time fishermen make up about 12 per cent of the total.

that has changed – the UK will almost certainly be leaving the CFP. This means the waters of the UK's 200 nm EEZ would come under national control for the first time and, at the same time, complete sovereignty would be regained over inshore waters. With property rights over the surrounding seas firmly vested with the UK's own national government,[52] fisheries management could then be carried out according to the long-term interest of UK nationals, taking on board the lessons learnt from the CFP and fisheries management systems in other parts of the world.

The UK government and the devolved authorities already have in place their own detailed system for allocating national quota among competing UK fishermen (see Appendix). They also have a system of policing these harvesting rights and for quite rigorously regulating the capacity of the industry. In the short term, this system could be continued, the crucial difference being that overall national quotas for each commercial fish stock in UK waters would now be determined by the UK itself, based on best national and international scientific advice, rather than bargaining between the EU fisheries ministers. It would also be purely national (rather than essentially European Commission) policy to determine the best conservation measures to use.

Unless the UK negotiates to stay within the EU Single Market, it will be free to decide on what terms (if at all) it wishes to continue to allow fishing businesses owned by

---

52 UNCLOS III gives coastal states the sovereign right to govern their EEZ.

non-UK nationals, but based in the UK, to have access to UK harvesting rights. At present, for instance, 43 per cent of the English fishing quota is held by foreign-owned businesses and 23 per cent is allocated to just one giant Dutch-owned fishing vessel, the *Cornelis Vrolijk*, which lands its entire catch in the Netherlands (Greenpeace 2015).[53] The UK government might also consider the degree to which it continues to allow boats based in EU Member States to claim historic rights to fish in the 6–12 nm zone of territorial waters and which, under the CFP, have been able to ignore British inshore conservation measures. These rights might now be the subject of negotiation, particularly as the UK at present has few similar rights to fish in the inshore waters of other EU Member States,[54] or they might be abandoned entirely simply through withdrawal from the 1964 European Fisheries Convention.[55]

More significantly, any decision to allow foreign vessels access to fish in the wider 200 nm zone would be a matter for bilateral negotiations and agreement between the UK

53 In 2015, a super trawler owned by the group owning the *Cornelis Vrolijk*, the *Frank Bonefas*, was caught by one of the Royal Navy's Fisheries Protection Squadron with an astounding 1,400,000 lbs of mackerel on board, all of which had been caught in a protected area off the Cornish coast. Despite the catch being worth up to £750,000, the owners were fined just £97,000, and allowed to sell their catch (see http://britishseafishing.co.uk).

54 For example, French vessels have the right to access no fewer than 15 different areas within the UK's coastal waters, whereas British vessels can only access one area within French coastal waters (Fresh Start 2013: 91).

55 Some legal experts argue that this is not necessary since the Convention has been superseded by EU law, so that when the UK leaves the EU these rights will automatically lapse (House of Lords 2016: 16).

and other coastal states. As Professor Robin Churchill has stated (House of Lords 2016: 15):

> [I]f in a particular coastal state's EEZ the coastal state is capable of harvesting the entire allowable catch, it is under no obligation to allow any other fishermen from other states to fish there, so it can take the whole of the allowable catch. Where an obligation to admit other fishers comes in ... that is where the coastal state does not take the whole of the allowable catch and there is a surplus. It must admit other states to the surplus, but again it has a discretion.

Management of fisheries could be conducted at the most appropriate ecological unit for the fish stock concerned: probably on the basis of sea basins (such as the North Sea, the Irish Sea, the Celtic Sea, the Channel, etc.) for most demersal species, and larger areas for migratory pelagic fish, with the UK entering into bilateral arrangements over fish management and conservation with the EU or other nation states as appropriate, as Norway, Iceland, Greenland and the Faroe Islands do at present.[56] Indeed, the majority of the UK's negotiations will, in future, be with these northern nations. Cutting out the Brussels bureaucracy should allow the UK fishing authorities to reduce red tape and act more speedily in response to the changing conditions surrounding fish stocks.

---

56 Norway, for instance, shares 90 per cent of its fisheries' stocks with other nations, so national quotas are set in cooperation with Russia, Iceland, the Faroe Islands, Greenland and the EU.

Over the longer term, the UK might follow some of the examples discussed elsewhere in the book and make the quota allocated to individual fishing vessels more fully tradable than it is at present. Alternatively, for sea areas with highly mixed fishing (such as the North Sea), the UK might abandon the quota system altogether in favour of 'days-at-sea' effort controls, as advocated by some groups of fishermen (Fishing for Leave/Save Britain's Fish 2017). Certainly, as a result of all these changes, fishing opportunities for the UK fleet should increase and subsidies will no longer be needed to encourage the decommissioning of fishing boats.

Appendix: UK system for apportioning national fishing quotas[57]

The UK divides the national quota it is allocated for each fish stock subject to TACs between *groups* of licensed fishing vessels largely on the basis of fixed quota allocation (FQA) units. These are abstract units of measurement based on vessels' historic share of national landings of this fish stock, usually for the period 1994–96. Essentially they are a right to harvest fish. For vessels over 10 metres long, these FQAs are assigned to individual vessels' licences; for those under 10 metres they are held as a block by the four fisheries administrations (see below). The FQA units are not fixed allocations of quota to the vessel in question, they are used as a mechanism for allocating quota.

---

57 This is very complicated and there are some annual variations in the way the different fish stocks are allocated (see DEFRA 2016).

The UK government first divides the quota for each fish stock between the four devolved fisheries administrations (FAs): DEFRA/The Marine Management Organisation (England), Marine Scotland, The Welsh Government and the Department of Agriculture and Rural Development (Northern Ireland). This is largely on the basis of the share of the UK FQA units held by the vessels registered with each of the FAs. Each FA has discretion as to how it allocates its share of the quota, but for England it is roughly as follows:

1. The total quota is apportioned between three groups:
   (a) 'The Sector' (vessels that are members of one of the 23 UK Producer Organisations).
   (b) The non-Sector pool (vessels over 10 metres that are not members of, or assigned to, a PO).
   (c) The 10-metres-and-under pool (the 'inshore fleet', vessels under 10 metres that are not members of a PO).

   For groups (a) and (b), this apportionment is on the basis of the FQA units assigned to vessels in the group; for group (c) it is based on the relative proportion of landings by 10-metres-and-under pool fishing vessels of each nationality in the period 2008–12. About 95 per cent of the UK's fishing quota is held by the Sector. Because of concern about the need to sustain the 78 per cent of the UK fishing vessels that make up the inshore fleet, are vital for local communities and which also practice the most sustainable fishing, there is now an 'underpinning' arrangement to top up the quota allocation of the non-Sector and the 10-metres-and-under fleet to a guaranteed minimum level. Many consider this to be inadequate and think that the underpinning arrangements need to be amended so as to provide more of the quota to smaller vessels.

2. The management of quotas within these three groups is as follows:

   (a) POs are responsible for managing their own quota allocations and making sure they are not exceeded. Some set monthly catch limits; others issue annual vessel or company quotas.

   (b) Quota allocations for the non-Sector pool and the 10-metres-and-under fleet are managed by the fisheries administrations. Each vessel's licence sets out the stocks that the vessel is not permitted to fish. For the non-POs, it also sets out monthly catch limits for the stocks the vessel is able to fish and land, which may be varied during the year as the national quota limit is reached. Apart from fish stock under particular pressure, where monthly catch limits may also be set, individual vessels in the under-10-metre fleet are generally allowed to fish without restriction until the overall quota allocation for the group has been taken in full, but this may be varied within the year.

Very limited markets operate within this system:

1. Since, in order to control the size of the UK fleet, no new fishing licences are currently created, in order to licence a vessel for the first time, an old licence (referred to as a 'licence entitlement') sufficient to cover the size and power of the boat, and the type of fishing required, has to be bought from previous licence holders removing their vessels from the fishing fleet.

2. The FQA units attached to old licences may be traded separately.

3. Subject to various rules, some annual quota swapping, or 'quota leasing', can take place.[58] The UK as a whole can swap quotas with another EU member state. The FAs can also swap quotas between themselves and with other EU member states, as well as negotiate quota swaps for all three groups between themselves, with the other two groups, or with another EU member state.

References

Ashworth, J. (2016) *The Betrayal of Britain's Fishing.* Worksop: Campaign for an Independent Britain.

Booker, C. and North, R. (1996) *The Castle of Lies.* London: Duckworth.

Booker, C. and North, R. (2005) *The Great Deception.* London: Continuum.

---

58 Over the period 2005–8, EU member states swapped, on average, over 10 per cent of their quota with each other (Fresh Start 2013).

Churchill, R. and Owen, D. (2010) *The EEC Common Fisheries Policy.* Oxford University Press.

COM (1966) Report on the situation in the fisheries sector of EEC member states and the basic principles for a common policy. COM(66) 250 (in French).

COM (1991) Report from the Commission to the Council and European Parliament on the Common Fisheries Policy. SEC (91) 2288, European Commission.

COM (2001) Green Paper on the future of the Common Fisheries Policy. COM/2001/0135, European Commission.

COM (2002) Communication from the Commission on the reform of the Common Fisheries Policy. COM/2002/0181, European Commission.

COM (2008) Reflections on further reform of the Common Fisheries Policy. Working Document, European Commission.

COM (2009a) Reform of the Common Fisheries Policy. Green Paper COM(2009) 163, European Commission.

COM (2009b) An analysis of existing rights based management (RBM) instruments in member states and on setting up best practices in the EU. FISH/2007/03, European Commission.

COM (2014a) Communication from the Commission to the European Parliament and the Council. Concerning a consultation on the fishing opportunities for 2015 under the Common Fisheries Policy. COM(2014) 388, European Commission.

COM (2014b) Seventh Annual Report of the Implementation of the European Fisheries Fund. COM(2014) 738, European Commission.

COM (2015) Communication from the Commission to the European Parliament and the Council. Consultation on the fishing opportunities for 2016 under the Common Fisheries Policy. COM(2015) 239. European Commission

COM and Parliament (2013) Regulation (EU) 1380/2013.

Council (1976) Council Regulation (EEC) 101/76.

Council (1983a) Council Regulation (EEC) 170/83.

Council (1983b) Council Regulation (EEC) 2908/83.

Council (1986) Council Regulation (EEC) 4028/86.

Council (1990) Council Regulation (EEC) 3944/90.

Council (1992) Regulation (EEC) 3760/92.

Council (2002) Regulation (EC) 2371/2002.

Council (2009) Regulation (EC) 1224/2009.

Council and Parliament (2014) Regulation (EU) 1380/2013 of 11 December 2013.

DEFRA (2016) *Rules for the Management of the UK's Fisheries Quotas.* London: DEFRA.

El-Agraa, A. (2011) *The European Union.* Cambridge University Press.

European Court of Auditors (2007) Special Report No. 7/2007.

European Court of Auditors (2011) Special Report No. 12/2011.

Fishing for Leave/Save Britain's Fish (2017) *The Brexit Textbook on Fisheries.*

Fresh Start (2013) *Common Fisheries Policy.* Fresh Start Project (http://www.eufreshstart.co.uk/common-fisheries-policy).

Greenpeace (2015) Our net gain (http://www.greenpeace.org .uk/media/press-releases/government-answer-legal-chal lenge-over-'unfair'-uk-fishing-quota-20150424).

HM Government (2014) Review of the Balance of Competences between the United Kingdom and the European Union, Fisheries Report.

House of Commons (2010–12) *Environment, Food and Rural Affairs Committee.* EU proposals for reform of the Common Fisheries Policy, Volumes 1–3.

House of Lords (2008) The progress of the Common Fisheries Policy. HL Paper 146-1.

House of Lords (2016) European Union Committee. Brexit: Fisheries.

Lequesne, C. (2000) Quota hopping: the Common Fisheries Policy between states and markets. *Journal of Common Market Studies* 38: 779–93.

Marine Management Organisation (MMO) (2015) UK Sea Fisheries Statistics 2014.

Minford, P. and Shackleton, J. R. (2016) *Breaking Up Is Hard to Do: Britain and Europe's Dysfunctional Relationship.* London: Institute of Economic Affairs.

New Economics Foundation (2011) *Money Overboard* (http://www.neweconomics.org/publications/entry/money-over board).

Penas Lado, E. (2016) *The Common Fisheries Policy: The Quest for Sustainability.* Wiley Blackwell.

Poseidon Aquatic Resource Management (2010) FIFG 2000–2006 Shadow Evaluation. Final Report for Pew Environment Group.

Wise, M. (1984) *The Common Fisheries Policy of the European Community.* London: Methuen.

WWF (2013) *Recovery of European Fish Stocks and the Reform of the Common Fisheries Policy.* Hamburg: World Wildlife Fund.

4 GOVERNING THE FISHERIES: INSIGHTS FROM ELINOR OSTROM'S WORK

Paul Dragos Aligica and Ion Sterpan

Introduction

The conventional approach to overfishing is for government to impose a top-down management system on the industry. This may involve a state agency setting overall catch quotas and then allocating shares to fishing enterprises, or creating some form of market framework that allows catch shares to be traded. Policymakers have tended to assume that fisheries will inevitably suffer from the tragedy of the commons with individual fishermen strongly incentivised to overexploit the resource and that only state intervention can prevent it.

The research of Elinor Ostrom[1] questioned this perspective. It found that many local communities around the world have evolved their own approaches to managing fisheries without the need for government intervention (Ostrom et al. 2012). They set their own rules on who has access to the resource, how it can be fished and

---

1   Elinor Ostrom (1933–2012) won the Nobel Memorial Prize in Economic Sciences in 2009, for her analysis of economic governance, especially of 'common-pool resources' such as fisheries (see Ostrom et al. 2012).

what sanctions will be imposed if violations occur. Such management models have often been highly effective at conserving stocks and maintaining yields in the long term, in marked contrast to the failure so often observed under state regulation of the sector.

It is clear from Ostrom's work, however, that there is no correct way to manage fisheries that will always be effective. Different models will be appropriate in different contexts, depending, for example, on local cultural norms and the physical characteristics of a fishery (ibid.). It is also accepted that in some circumstances there may be a limited role for the state, though more in the sense of providing a stable legal framework to assist enforcement processes than the imposition of prescriptive regulation.

This chapter examines the practices communities have adopted to manage their fisheries and discusses the conditions necessary to make such community-based approaches both possible and effective. To evaluate the performance of alternative models we need to look at the choices that fishermen face in their particular environments and at the institutional arrangements framing those choices. The importance of context raises doubts regarding the viability of broad and universal solutions derived purely from general principles (see Ostrom 1975). While specific models can make an important contribution to tackling overfishing, they are only suitable for particular locations under suitable conditions. In other words, there is no one-size-fits-all solution to the problems facing the world's fisheries.

Public choice and voluntary action

Elinor Ostrom's focus on institutions, as well as the ensuing challenge to think beyond the simple models of 'the state' and 'the market', can lead to a misunderstanding of her views. The idea of going beyond the 'market–state' dichotomy may generate uneasiness among pro-market authors used to 'private versus public' framing. Yet, Ostrom's stance may be seen as compatible with the classical liberal perspective. As Pennington (2012) explains,

> Ostrom and classical liberals argue that there is a significant class of environmental problems, including the management of forests, watersheds, inshore fisheries and many local collective or public goods where it would be better to rely on more decentralised forms of management. For classical liberals, by decentralising decision-making to a variety of individuals and organisations a private property system facilitates a greater level of experimentation than more state-centric regimes, allowing for emulative learning while minimising the impact of inevitable mistakes.

The idea is simple: individuals are confronted with problems and they need to cooperate to solve them. They may try markets, or they may decide voluntarily to transfer certain decision-making powers to the level of the community. They may also voluntarily search for more inclusive cooperative arrangements that go beyond the market or community-based mechanisms.

Ostrom's approach, to use the words of James Buchanan (1964: 222),

> concentrates attention on the institutions, the relationships among individuals as they participate in voluntary organised activity, in trade or exchange, broadly considered. People may decide to do things collectively. Or they may not. The analysis, as such, is neutral in respect to the proper private sector–public sector mix.

In brief, the Ostrom approach is based on the assumption that we need to concentrate on voluntary actions and the processes and institutions that emerge from them. A certain scepticism regarding coercion is also presumed. At the same time, the perspective assumes basic common sense. It does not claim that one type of institutional arrangement is optimal in all circumstances, irrespective of natural, social, cultural and technological factors, and irrespective of the preferences of the people and communities involved.

The approach

So far we have emphasised the centrality of the principle of voluntary choice and the diversity of institutional arrangements that include but go beyond market and state. Let us take a further step by introducing two aspects that are central to Ostrom's approach to the problems of governance in circumstances (such as those of fisheries) in which a commons element is central or salient.

The first is strongly empirical. Field work has shown that the conceptual models traditionally used to understand 'the tragedy of the commons' are limited in application. Ostrom refuted the statement that all commons represent problems that are insoluble without intervention by a state-like institution. She made the point that traditional models, such as the tragedy of open-access regimes that Hardin thought applied to all commons (Hardin 1968),[2] only describe special – limited and limiting – cases in a range of commons situations. In real-world settings, individuals have in many cases the resources to change the rules of the game and overcome the tragedies and dilemmas. Ostrom inspired an entire line of research documenting this.[3]

The second aspect is mostly analytical. Ostrom developed a structure for the purpose of describing, analysing and understanding institutional settings. Alongside real-world examples, this maps possible ways out of social dilemmas and collective action problems. Ostrom's framework draws attention to the fact that those ways go beyond the binary 'market–government' solution space.

---

2   Also, the size-logic of collective action that Olson thought applied to all real-world situations (Olson 1965) and the multiplayer prisoner's dilemma game (Dawes 1973).

3   For example, Pomeroy et al. (2001) document success in co-managed Asian fisheries, while Gutierrez et al. (2011) analyse over 130 co-managed fisheries worldwide and find that success is highly correlated with strong autonomous involvement by local communities. The Seri pen shell fishery (Basurto 2010) and the lobster fishing community in the Gulf of Maine (Acheson 2003) provide detailed success stories of community-based management (see below).

The challenge is to overcome the presumption that any resource not privately owned can only be managed either by the government or left with open access.

Governance regimes

At the most simple level the framework of analysis starts from a typology of four governance regimes: open access, private property, state property and common property. Each of the last three regimes solves in some manner the problem of resource depletion, i.e. the tragedy of open access. Ostrom breaks down the problem of depletion into three subproblems: (a) the problems of institutional supply, (b) the problem of commitment and (c) the problem of enforcement (1990: 42–47). The three problems can be formulated as questions. Who pays the cost of defining the rules and how? How do individuals who commit to follow the rules acquire credibility? Who pays the enforcement costs and how? Significant insights may be gained by simply asking and trying to answer these simple questions.

Starting from empirical studies of resource systems such as fisheries, water basins, forests and irrigation schemes, Ostrom tried to identify the conditions in which each type of management regime has a relative advantage over the others. When showing how members of a commons situation solve the problem of depletion, Ostrom offers an account of successful collective action in the supply and enforcement of specific sets of rules. The aim is not to show that a certain regime (for instance, collective property of fisheries) is an optimal universal arrangement that should in principle be preferred to the others. Instead, the goal is to discover the circumstances in which a regime satisfies certain governance or economic performance criteria. That is to say, to determine when and where a governance and property regime is more effective and to explain why.

As already mentioned, the basic observation is that in many cases, communities may be able to extract themselves from the non-cooperation characteristic of the tragedy of the commons (Ostrom 1990: 48). Under certain conditions, users can create functional cooperation arrangements. For instance, the capacity to exclude outsiders from using a resource matters. Success can often only be achieved when outsiders, the central government among them, can be denied access.

Ostrom does not claim to offer a list of sufficient conditions. The effort of building recipes is plagued by the problem of translating local, implicit and transitory knowledge into explicit knowledge (Agrawal 2001). However, those institutional arrangements that work do so because they draw on, closely match and link well to local norms, local culture and local particulars, including the physical attributes of the resource.

This is the context in which one should see Ostrom's observation that under certain circumstances, group governance is more effective than both larger government units (states) and individual private owners linked by a market. On the one hand, the local self-governing group is a better owner relative to the larger government due to the knowledge advantage of local users, combined with inexpensive rule enforcement, which is provided spontaneously when members naturally monitor each other in their daily activities. On the other hand, the self-organising group may be a better manager than separate individuals linked by markets when the technical difficulty of dividing the resource requires government protection of private property rights. The latter may be an expensive option compared with enforcement of rules within the group by members.

Performance criteria

The diversity of criteria for evaluating the performance of a governance regime looms large in Ostrom's perspective. Nevertheless, the most commonly used criterion of success continues to be centred on harvest maximisation or maximum sustainable yield of the particular resource in the long run (Hilborn 2007: 297). The alternative criteria fall into three classes: social, political and economic. Social objectives may include the spread of employment and income, as well as maintaining traditional communities. Political goals include maintaining demand levels and employment (hence the subsidies examined in Chapter 2).

Assessment should not be focused exclusively on the local community, disregarding the effects on neighbouring economies and communities. All criteria of performance are potentially non-local in outlook. Property regimes are called upon to solve the problem of depletion where open access would be mostly subtractive, all neighbourhood

activities considered. Economic criteria dictate, for example, that a fishery should move away from open access
when the total rent from appropriation is diminishing, not
only dissipating among more numerous users, regardless
where they come from. But similarly, political criteria can
also take into consideration aggregate values of demand
and employment, values not bound to a single community,
activity or sector. Even concerns for maintaining the traditional character of communities of fishermen may be
expressed by actors and groups that are outside the group
or community in question.

In some situations and cases, different types of criteria
(biophysical, social, political and economic) overlap, and
a simple intuition of sustainability is enough. In other
cases, criteria conflict. For instance, biological success
may not generate an economically healthy fishing community when strong conservation measures imposed to
try to rebuild an overfished stock of fish A cost too much
in forgone yield from healthy stocks of fish B, which shares
the same habitat. This kind of problem partly explains why
successful autonomous fishing communities try to avoid
quotas (Wilson et al. 1994: 304). The loss in yield due to governments indirectly restricting quotas of healthy stocks of
fish B to avoid by-catch of overfished stocks of fish A, can
sometimes be massive (Hilborn 2007: 299). If weak governments are unable to move resources away from open
access conditions (overfishing), underfishing is the trap
threatening stronger governments of wealthier countries
that overlook precisely that kind of loss.

Co-management and the commons

One of the main questions raised by Elinor Ostrom's work is: what are the conditions and variables that signal community governance may be a feasible and desirable solution? Ultimately, explains Ostrom (McKean and Ostrom 1995: 6),

> Common property regimes are a way of privatizing the rights to something without dividing it into pieces... Historically common property regimes have evolved in places where the demand on a resource is too great to tolerate open access, so property rights have to be created, but some other factor makes it impossible or undesirable to parcel the resource itself.

The argument in favour of community-based projects has robust empirical support when it comes to fisheries. One thread running through all successful and profitable fisheries is some form of secure and limited tenure or access to the fishery (Hilborn 2007). The idea of communal tenure and co-management is strengthened by another related Ostrom contribution, polycentricity.

Decentralised, polycentric systems that feature mutually adjusting small providers with fragmented jurisdictions and overlapping areas of focus provide better governance than larger, centralised governments. Co-management, authority-sharing between levels of government, and resource sharing schemes are unavoidable when there are many complex problems transcending the boundaries of traditional jurisdictions (Espectato et al. 2012: 27). For example, industry, regulating agency and legislature successfully cooperate in the Gulf of Maine lobster fishery (Wilson et al. 1994; Acheson 2003).

Government interference and failure

An important part of Ostrom's work was dedicated to identifying 'design' principles for successful community management (1990: 90). This creates the possibility of a more systematic analysis to identify the hindrances to institutional arrangements becoming robust. Cases in which government interference hinders rather than helps the solution to a problem are of particular interest. As Ostrom (2000: 138) put it: 'Solid empirical evidence is mounting that governmental policy can frustrate, rather than facilitate, the private provision of public goods'. In the context of a discussion of fisheries governance, it is important to consider this theme.

For instance, the first principle (1) and the foremost condition of communal ownership is closed access, the ability of the group to control access to their fishing grounds (Hilborn 2007). Compared to other common pool resource systems, fisheries are particularly vulnerable in terms of defining the boundaries and the number of individual appropriators. But meeting this principle is not impossible. The literature offers many examples of success. The lobster fishing community in the Gulf of Maine successfully limits entry (see below). The people of the successful Seri pen shell fishery have rules which grant access to and authorise outsiders (Basurto 2010). Authorisation is not a ratchet-like mechanism that inevitably moves the number of appropriators towards overcrowding and overcapitalisation. Authorisation and licensing are reversible: when uproar about rule breaking is created within the community (whether justifiable or not), the permits to authorised entrants are forfeited despite their protests. Analysing the failed and fragile Bodrum, Izmir, Nova Scotia and the Sri Lankan Mawele examples discussed by Ostrom (none of which were able to control the number of appropriators) one finds government interference as a significant reason for failure.

The other institutional design principles, to summarise, suggest (2) congruence between appropriation rules and local conditions; (3) that most individuals participate in modifying operational rules (i.e. are present in decision-making arenas); (4) monitors be accountable to the appropriators; (5) sanctions be proportional to the infraction; (6) participants have low-cost access to conflict-resolution mechanisms; (7) participants enjoy recognition of rights to organise; and (8) activities be organised in layers of nested enterprises, such that levels of rules are mutually coherent. All of the above are vulnerable to distortion and undermining by government interventions. At the same time, understanding the limits and dangers of state intervention illuminates the proper sphere of constructive government action.

That being said, it is important to guard against the tendency to identify the state as the only root of difficulties. Ostrom's work follows the empirical reality on a case-by-case basis, taking into account all relevant factors. For example, a diminishing size of individual shares is not always due to outsiders whose entry rights are backed by actors outside the group, such as governments and their agents. The coming of new generations of fishers that inherit the rights of established families may destabilise the system. Ostrom (1990: 153) cites Paul Alexander: 'if there are twenty nets [and 8 shares in a net], a man with one share receives 1/160th of the annual catch, whereas if one joins in the construction of a new net, they each receive 1/168th'. Poorer economic performance may thus be the result of government decisions but also of other factors, both external and internal to the community.

The following two case studies illustrate the kind of community-based practices that have emerged to protect fish stocks and examine how they fit into the analytical framework developed by Ostrom. The impact of government policy on the two systems is then discussed. The Maine lobster industry in the US and the fishing industry in the Canadian provinces Nova Scotia and Newfoundland are located on the eastern seaboard of North America but experience different treatment by the relevant federal authority.

A case of fragile institutions: the Nova Scotian inshore fisheries

Over the generations, fishermen in Nova Scotia and Newfoundland have developed rules to govern the use of inshore resources. Depending on the time of the year, different fishing technologies are used to fish cod, halibut, herring, mackerel and lobster, the latter alone yielding up to 40 per cent of a typical fisherman's annual income.

Communities such as Port Lameron Harbor, described in detail by Davis (1984), are organised in villages where up to a hundred fishermen are on the water year-round in boats with an average crew size of two. These fisheries fulfil the conditions described by Ostrom that should be conducive to community-based management. In particular, fishermen share very strong social ties, typically as part of families with long traditions of fishing in that particular locality. Such community cohesion reduces the chances of insiders breaking the rules, since they fear social ostracism. At the same time, it facilitates collective action to prevent outsiders from encroaching on claimed territory. Davis describes how this works (Davis 1984: 147):

> [A] Port Lameron Harbor fisherman, after setting his longline gear, watched a fisherman from a neighboring harbor set his gear close to and, on occasion, across his line. Subsequently, the Port Lameron Harbor fisherman contacted the transgressor on the citizen band radio to complain about this behavior. Other Port Lameron fishermen who were listening in on the exchange demonstrated support for their compatriot by adding approving remarks once the original conversation had ended. The weight of this support, coupled with the implied threat of action, i.e. cutting off the offender's gear, compelled the erring fisherman to offer his apologies.

Fishing grounds are organised in zones extending both outwards and along the coast from each village for about 20 kilometres. Conflicting technologies – i.e. fishing methods that get in other fishermen's way – together with other factors have led to a division of each zone into sub-zones. Because each zone is matched to a suitable technology, divisions reduce the costs of monitoring violations of rules on how to fish. Moreover, matching each zone to a particular type of technology suited to its local environment minimises externalities and reduces costs from incompatibilities in types of gear. Zoning is also one method of ensuring a just distribution of the yield among fishermen (Ostrom 1990: 173–74).

The rule system is held together by Ostrom's first design principle: that of enforcing closed access. What makes the rule system fragile, though, is missing recognition by Canadian federal authorities, through the Department of Fisheries and Oceans (ibid.: 175).

Before 1949, when Newfoundland was incorporated into the Canadian confederation, the stance taken by the authorities had been to provide arenas for conflict resolution and to codify into formal law whatever informal rules were long-established within local communities. Today, fishermen still base their communal exercise of the right to refuse access to outsiders on tenure and on the memory that authorities strengthen locally evolved rules, especially when those rules have proved effective.

It is not hard to see that the missing recognition of closed access by federal authorities increases the cost of conflict resolution within communities of fishermen.

Under government sanctioned open access and with the burden of rule enforcement shifted entirely to the fishing community, the first half of the 1970s witnessed several instances of stock depletion (ibid.: 176).

Subsequent government interventions led to further adverse consequences. Animated by an otherwise shared goal of limiting resource use, in the late 1970s the federal authority introduced comprehensive licensing of vessels and gear. It then, without prior notice, froze the number of licences available while threatening sanctions for the use of unlicensed nets. Within that interval, in expectation of the second component of the policy, fishermen made costly efforts to obtain licences for gear they were not actually using, in order to insure against a time when they might actually need it. In the same period, driven by the goal of increasing yields but trumping locally developed strategies of resource use, the government subsidised local purchases of offshore groundfish gill nets. Their use interfered with inshore operations to the point that all beneficiaries eventually had to dispose of the new gear (Ostrom 1990: 174 citing Davis 1984). The interventions of the government proved to be environmentally damaging and economically wasteful, as well as undermining key aspects of the community management model.

A case of robust institutions: the Maine lobster industry

The Maine lobster industry accounts for up to 60 per cent of US lobster landings, worth up to US$186 million per year.

It is one of the world's most successful fisheries (Acheson 2003: 13, 206) with local communities playing a major role in its management.

Like its Canadian counterpart, Maine is a collection of ecosystems developed around small harbour communities of between 8 and 50 full-time fishermen and extending out to a median 6–7 mile radius. With the exception of political affairs carried out mostly by elected leaders, contact between harbour groups is kept to a minimum.

There is no record of how the territorial rules came into being (ibid.: 41), but evolved rules governing entry into harbour groups, placing limits on the number of traps used and governing the movement of border lines appear to be effective at minimising violence and maintaining sustainable yields.

These rules are at odds with simplistic preconceptions on the coherence, homogeneity and all-explicit character of the law. The laws of Maine, stating that ocean waters are public property, held in trust by the state and open to anyone purchasing a licence, officially remain in force. At the same time, public administration coexists with informal admission rules to territorial harbour groups and effectively allows closure around harbours. Without such coexistence, it could have proved difficult to restrict access to the fisheries.

Local rules have not always been nested in state laws. The 1920s saw rampant violations of ill-conceived state conservation laws, leading to a disastrous decline in yields (ibid.: 86). It took impressive political entrepreneurship by the Commissioner of Sea and Shore Fisheries, Horatio Crie, to break a long-standing unproductive stalemate between industry and government and among industry factions. On the one side stood an alliance between western fishing counties and dealers, who desired a lowering of the legal threshold size to allow them to compete with Canada in the small-sized lobster market. On the other side stood the eastern and central counties who were unfavourable to the small gauge because they feared it would allow dealers to import small lobster from Canada. The imported lobster would have undercut local prices.

In December 1933 commissioner Crie negotiated a deal in which he fulfilled the western faction's wish of lowering the minimum size limit in exchange for its support for a second, maximum threshold, meant to protect very large lobsters with high reproductive potential. The 'double gauge law' granted an immediate advantage to the western faction but laid the foundations for higher yields for the otherwise more numerous eastern and central counties. It remains a major component of a successful political compact between the state government and fishermen's local councils. The agreement is tuned and adapted in impartial proceedings and publicly provided arenas – for instance, chaired by university professors and held in local schools.

There is still some disagreement between scientists and local lore on the causes of the bust in the 1920s and 1930s. But awareness of this history may well play a positive role in generating and renewing a shared, overarching conservation ethic. It is indisputable that following the bust support among fishermen for imposing limits on lobster trapping for themselves steeply and then steadily increased, to the point that government support and sanction sometimes became unnecessary.

For example, after several uniform state-wide trap-limit bills had been defeated in the legislature in the 1960s and 1970s, various island communities relied on their small size, social cohesion and high dependence on the resource to self-impose conservation rules. These rules are strict but not uniform. Trap number limits typically vary from 400 to 600 per boat, with some flexibility tolerated to maintain good relations within the group. At the time Acheson conducted the research, a small number of fishermen in one location were not abiding by the 600 limit, but were escaping sanction because it was recognised that cutting their gear would result in an irreparable fissure within the harbour gang. In another community defectors escaped punishment because they were family members (ibid.: 63). Imperfections notwithstanding, the micro level of organisation in these islands is strong enough to demonstrate the rule generating and enforcing capability of fishing communities acting independently of government.

Conclusions

The main underlying message of Ostrom's approach is the simple but powerful notion that, when it comes to governance arrangements, one has to analyse each case on its merits, based on the relevant evidence. For instance, when it comes to fisheries, given their physical features, there are some general principles that may be used for guidance, but the specifics of the fishery in question, in terms of its social, ecological and institutional environment, should be the main driver. There is no one-size-fits-all solution in fisheries governance.

Ostrom's attitude is one of scepticism regarding 'the state' as a universal solution to governance problems. Indeed, in many cases the state may be a problem maker and not a problem solver. At the same time, Ostrom is convinced that there is overwhelming evidence that 'individuals in all walks of life and all parts of the world' are able to 'voluntarily organize themselves so as to gain the benefits of trade, to provide mutual protection against risk, and to create and enforce rules that protect natural resources' (2000: 138). Her message is not driven by doctrinaire assumptions but by analytical and empirical evidence. In this respect, her work represents an important contribution to policy analysis and public administration. Her theoretical lenses, principles and insights should be considered an essential tool in the management and governance of fisheries.

References

Acheson, J. (2003) *Capturing the Commons: Devising Institutions to Manage the Maine Lobster Industry.* Hanover and London: University Press of New England.

Agrawal, A. (2001) Common property institutions and sustainable governance of resources. *World Development* 29: 1649–72.

Basurto, X. and Coleman, E. (2010) Institutional and ecological interplay for successful self-governance of community-based fisheries. *Ecological Economics* 69: 1094–1103.

Buchanan, J. M. (1964) What should economists do? *Southern Economic Journal* 30(3): 213–22.

Davis, A. (1984) Property rights and access management in the small boat fisher: a case study from south-west Nova Scotia. In *Atlantic Fisheries and Coastal Communities: Fisheries Decision-Making Case Studies* (ed. C. Lamson and A. J. Hanson), pp. 133–64. Halifax: Dalhousie Ocean Studies Programme.

Dawes, R. M. (1973) The commons dilemma game: an *n*-person mixed motive with a dominating strategy for defection. *ORI Research Bulletin* 13: 1–12.

Espectato, L. N., Baylon, C. C., Serofia, G. D. and Subade, R. F. (2012) Emerging fisheries co-management arrangement in Panay Gulf, Southern Iloilo, Philippines. *Ocean & Coastal Management* 55(9): 27–35.

Gutierrez, N., Hilborn, R. and Defeo, O. (2011) Leadership, social capital and incentives promote successful fisheries. *Nature* 470: 386–89.

Hardin, G. (1968) The tragedy of the commons. *Science* 162: 1243–48.

Hilborn, R. (2007) Moving to sustainability by learning from successful fisheries. *Ambio* 36(4): 296–303.

McKean, M. and Ostrom, E. (1995) Common property regimes in the forest: just a relic from the past? *Unasylva* 46(180): 3–15.

Olson, M. (1965) *The Logic of Collective Action. Public Goods and the Theory of Groups.* Cambridge, MA: Harvard University Press.

Ostrom, E. (1990) *Governing the Commons. The Evolution of Institutions for Collective Action.* Cambridge University Press.

Ostrom, E. (2000) Collective action and the evolution of social norms. *Journal of Economic Perspectives* 14(3): 137–58.

Ostrom, E., Chang, C., Pennington, M. and Tarko, V. (2012) *The Future of the Commons: Beyond Market Failure and Government Regulation.* London: Institute of Economic Affairs.

Ostrom, V. (1975) Alternative approaches to the organization of public proprietary interests. *Natural Resources Journal* 15: 775–89.

Pennington, M. (2012) Elinor Ostrom on privatization. *IEA Blog* (http://www.iea.org.uk/blog/elinor-ostrom-on-privatisation -0).

Pomeroy, R., Katon, B. M. and Harkes, I. (2001) Conditions affecting the success of fisheries co-management: lessons from Asia. *Marine Policy* 25(3): 197–208.

Wilson, J., Acheson, J. M., Metcalfe, M. and Kleban, P. (1994) Chaos, complexity and community management of fisheries. *Marine Policy* 18(4): 291–305.

5 RIGHTS-BASED OCEAN FISHING IN ICELAND

Birgir Runolfsson

For most of the last century the world's fisheries were out-side the jurisdiction claimed by coastal nations and thus subject to pure open-access conditions, often referred to as 'common property'. With larger and more effective fishing fleets, coupled with the rise in demand for fish, rapid and dramatic overexploitation of fish stocks resulted. Fisheries management was limited and largely ineffective.

In the last decades of the twentieth century, countries shifted the management of ocean fisheries within 200 nautical miles of their coastline from open access to intensive regulation. Governments attempt to restrict the total harvest of fish in order to stabilise or increase fish stocks. Yet such regulatory regimes have largely failed to stem the decline of fisheries because they do not alter the fundamental incentives that lead to overfishing. Change is therefore inevitable in the fisheries. Managing a fishery through top-down regulation does not solve the basic incentive problems caused by the lack of property rights to the fish stock. Excessive fishing still exists because of the absence of property rights.

Recently, several countries have replaced fisheries managed by command-and-control regulations with systems based on property rights. Rights-based fishing is increasingly recognised as a practical alternative to the inefficiencies of direct controls and regulation. The role of property rights in fisheries should be no different from the role of property rights elsewhere in the economy: property rights, if adequately defined and enforced, encourage efficient use of resources in the present with an appropriate regard for the future. Fisheries are but the last of the 'commons' resources to which private property rights will develop. History tells of enclosures and clearances of common land in response to changed economic circumstances. The private property system for land and other resources is responsible for increases in economic productivity in recent history. The expansion of property rights as a method of economic organisation should extend to transferable harvesting rights in fisheries. As with property rights on land, rights-based fishing will yield substantial economic benefits.

The fisheries problem

Only a generation ago, the supply of fish available from the world's oceans seemed plentiful. However, advances in fishermen's ability to catch, preserve, transport and sell fish quickly exceeded the ability of fish stocks to reproduce. World marine catches increased more than fourfold from 1950 to 1990, from less than 20 million tonnes to more than 80 million tonnes, but have stagnated at that level since. Furthermore, most of the world's most valuable fish stocks are either fully exploited or overexploited, and, in economic terms, more than 75 per cent of the world's fisheries are underperforming or are subject to economic overfishing. Though most fisheries are biologically and technically capable of yielding high net economic returns on a sustainable basis, few actually do.

As a whole, the ocean fisheries appear to yield very small or even negative net economic returns. A study by the World Bank and FAO found that in 2004 the global ocean fishery operated at a significant net economic loss, an estimated $50 billion a year, and this loss was often financed by government subsidies. By contrast, the same study found that with proper management, the global fishery could yield positive net returns of more than 50 per cent of revenues on a sustainable basis (World Bank and FAO 2009).

Governments have responded to the decline in fish stocks with command-and-control regulation. These regulatory regimes attempt to reduce overfishing through various types of restrictions, including limits on the length of fishing seasons, the number of fishermen, vessel size, types of gear and the amount of fish that can be caught. Because such regulation rarely works, additional measures have been tried in order to limit the intensity of fishing efforts and number of fishers in a given fishery, including limits on investment in fishing efforts, buyback schemes, licensing and limited entry.

While such regulations drive up costs and discourage some fishing effort, they do not alter the fact that fish are valuable but no one owns them. Fish that are in the waters today may not be there tomorrow. Those who catch the fish earn money. This fact, as well as the existence in many countries of government subsidies for the acquisition of boats and gear, encourages fishermen to explore further means for finding fish. For example, limits on vessel size encourage investment in more boats and in more sophisticated equipment; specifying which days of the week, month or year one can fish encourages more intensive effort on those days. Restrictions on fishing efforts make fishing less efficient than it could be. Seasonal closures coupled with improved fishing technology most often results in overcapitalisation and wasteful racing for fish.

Overfishing and other inefficient fishing practices have nothing to do with the nature of the resource, the characteristics of fishermen or the localities in which fish are found. Rather, inefficiencies are the direct result of the definition and enforcement of property rights in fisheries, or rather the lack of these. Fisheries are troubled by overfishing because private property is lacking. Fishermen own only what they catch. The government, which is to say, everyone and therefore no one, owns the stock of fish from which the catch is taken.

Creating rights to fishing

If fish stocks were privately owned, incentives would exist to conserve them because the gains from their preservation as well as the costs of their exploitation would accrue to their owners. Private owners will neither race to take fish nor deplete stocks that would enhance future catch because if an owner does either, he bears the cost. The fisheries problem is therefore, in a sense, man-made. It stems from a particular social arrangement stipulating that everyone, or at least everyone belonging to a defined group, can harvest the fish stocks. The obvious remedy, therefore, is to replace this social arrangement with one – rights-based fishing – stipulating that only those with well-defined rights to harvest can fish (Neher et al. 1989; Scott 2008; Arnason and Runolfsson 2008). These rights, obviously, amount to private property rights which have been well-established as being efficient in other areas of economic production. There are several possible types of private property rights in fisheries (Arnason 2012; Wilen 2006; Wilen et al. 2012).

Common types of the less-than-perfect property rights used in ocean fisheries are territorial use rights (TURFs) and individual quotas (IQs) that may be transferable (ITQs). Under TURFs, fishers are allocated a certain area of the ocean, very much like a plot of land, for their exclusive use. The establishment of private ownership in coastal fisheries, where fish stay put, is conceptually simple and very analogous to private property on land. A coastline could be carved up and private owners would be allowed to take exclusive possession of the fish in their area, a TURF. The problem with this approach is that most species of fish (not to mention their eggs and larvae) move around so much that either the TURFs would have to be huge in order to enclose them or additional coordination mechanisms are required. As a consequence, TURFs are generally applied only to relatively sedentary species such as certain species of shellfish.

A system of Individual Transferable Quotas (ITQs) modifies simple Total Allowable Catch (TAC) regulations to prevent the race for fish. Under an ITQ system, the TAC is allocated as individual quotas to fishermen, fishing firms, or fishing vessels. ITQs are rights to harvest a certain volume of fish. While ITQs are more widely applicable than TURFs, they are not property rights in the resources themselves (i.e. the fish stocks and their ocean habitat). This limitation reduces the quality of ITQs as property rights and therefore their effectiveness in maximising the flow of economic benefits from the fishery.

An ITQ system giving operators a right to a share of the harvest is not as good as a right to all fish in a defined territory. ITQs are not perfect rights because the gains from behaviour that negatively affects the stock of fish, like cheating on one's quota, accrue to only one person, while the losses are dissipated among all other owners of the quota. But because ITQs provide security for a share of the harvest, fishermen will not dissipate the wealth in a fishery by competing among themselves for a greater share of the total catch. Even though ITQs are not ideal property rights, they provide a practical and politically achievable reform for the traditional ineffective systems of fisheries management.

After the initial quotas are set, fishermen are free to adjust their share by buying, selling, or leasing a quota. This approach allows fishermen to better respond to market conditions by adjusting the nature, timing and scale of operations to produce a more profitable harvest. The quotas in an ITQ system should be proportional (the right to a percentage of the TAC) and permanent property rights. Absolute changes in the TAC will then translate into proportionate changes in each individual's quota holdings without any adjustment in the ITQ. The ITQ should also be allocated in perpetuity. Fishermen with a permanent interest in the harvest will manage their behaviour more efficiently.

There has been a great increase in the use of ITQ systems in fisheries around the world in the past four decades since their introduction. That by itself suggests that they are generally found to produce at least some benefits compared to the alternatives. The particulars of individual quota systems do vary greatly, not least in the degree of quota tradability, and it can often be difficult to distinguish between systems of tradable and non-tradable quotas. Nevertheless, it can be safely asserted that ITQs have been adopted in hundreds of individual fisheries around the world. The first ITQ systems in ocean fisheries were introduced in the 1970s and by 2010 at least 22 significant fishing nations were using ITQ systems as a major component of their fisheries management. It has been estimated that the total volume of marine catch taken under ITQs may be about 22 million tonnes, out of the annual global ocean harvest from capture fisheries of just over 80 million tonnes in recent years. Catch taken under ITQs is therefore as much as a quarter of the global harvest (Arnason 2012).

Criticism and concerns about an ITQ system

Despite their growing acceptance, ITQ systems have attracted criticism. Several different claims are frequently made and we will examine the most important ones here.

The first criticism concerns our understanding of fish stocks and that it is insufficient to determine the correct TAC. The critics are correct that fisheries management is as much art as it is science. But the scientific limits of our knowledge of fishery dynamics affect all fisheries management systems equally. That is because the TAC concept is a central feature of all systems of fisheries management. The purpose, whether it is explicit or implicit, of the restrictions and regulations in all systems of fisheries management is to limit the catch to a level that a fishery can tolerate. The explicit TAC in an ITQ system is preferable to the indirect ineffective methods of limiting the catch.

The benefits of an ITQ system exist even in the presence of scientific uncertainty about the long-run sustainability of any particular TAC. The TAC will be continuously adjusted because of the inherent biological variability in fisheries and their ecological interrelationships. Our understanding of those issues, and hence our ability to set TAC at a sustainable level, should improve over time. Whether the TAC is set too high or too low will not affect the assertion that ITQs will maximise income from the TAC. For most fisheries, only a TAC that is set too high year after year will create difficulties. There is also some evidence that under ITQs the previous long-term decline of fish stocks has been halted and even reversed (Costello et al. 2008, 2010). This empirical evidence, limited though it may be, fits well with the economics behind ITQs.

Another criticism concerns the discarding of fish. Although discarding in world fisheries is well known and estimated to be quite high (World Bank and FAO 2009), there is concern that this problem is even larger under ITQs. When ITQs are used in a multi-species fishery there may be a problem of by-catch. That is, fishing vessels aiming to harvest particular species which they have quota for may harvest other fish for which they do not have quota and will therefore discard those fish. But ITQs are now being used in multi-species fisheries and the lessons learned from that experience indicate that this is not really a problem. Fishers most often have sufficient mix of quota for the various species that are likely to be by-catch.[1]

In addition to by-catch problems, critics claims ITQs encourage 'high-grading'. This refers the process of discarding smaller fish in the hope of catching larger, more valuable ones.[2] Providing proper incentives for fishers and sufficient monitoring of their actual behaviour should reduce high-grading and other discarding of fish.

---

1   One way to address the potential discard problem of by-catch is to have some flexibility in the system. Sufficient flexibility in balancing catches after the fact by acquiring additional quota for the by-catch could help. Another option to increase flexibility would be to establish 'equivalent rates' of fish species, whereby catch in one species can be covered by quota of another species. Yet another is to allow landings of some by-catch that would not be counted towards the fisher's quota, but where the fisher has to surrender the catch value of that by-catch.

2   This problem arises, at least partly, due to the fact that the quota refers not to the number of individual fish but rather to the weight of fish. For some species of fish market prices could be such that one kilo of 'big fish' is more valuable that two kilos of 'small fish'. This situation could provide incentives for 'high-grading'. How much of a problem this would be depends on the price dispersion between the different sizes, as well as the costs and benefits of additional fishing, transporting, etc. Monitoring and enforcement of discarding also matter of course, as well as the flexibility mentioned in the previous footnote.

The problems of high-grading and discarding also seem to be much smaller than claimed and the empirical evidence suggests, contrary to critics, that lower discard rates are one of the benefits of catch-share systems. A recent paper reports that the discards-to-retained-catch average in the fisheries studied actually fell by almost a third over a five-year period and two thirds over ten years. Almost all the fisheries studied reported a lower discard rate under catch shares than under traditional management (Grimm et al. 2012).

Yet another criticism is concentration of quotas: that communities or geographical regions may suffer quota loss and that 'smaller' fishermen will lose out to 'bigger' fishermen. The empirical evidence would appear to offer some support to this assertion. The reason is the improved economic efficiency of the fisheries, and from that viewpoint these results are to be welcomed. In fact, many would point out that a key purpose of reforms in the fisheries is to decrease the number of fishing vessels that are chasing the fish, and fewer vessels result in fewer fishing firms operating.

But if there are concerns about this there is always the option of limiting these effects. An ITQ system could limit ownership concentration through regulatory caps, set aside community quotas which may only trade within a community or region, and set up separate ITQ systems for 'small' and 'big' vessels that have more restricted trade between systems. Such limitations will no doubt come at the cost of economic efficiency and that should be acknowledged explicitly by policymakers.

A related criticism is the concern that harvesting rights will exclude new fishermen. The perception that closing the commons excludes some from access to fishing is accurate, but the concern is overstated. The fishing of ocean resources is currently excessive, so by definition, some who are currently fishing will not be fishing in the future. This fact is unaffected by the management system in place. The ITQ system, in fact, is superior to the traditional system because as long as people can trade the quota rights, nobody is automatically excluded. And once you obtain an ITQ right, the fish will actually exist for you to catch. Under a traditional system, everyone is free to fish, but the race to harvest often implies only a 'right' to harvest fish at no profit, a right worth nothing.

The argument that ITQs allow the use of fisheries by some people to the exclusion of others often seems nothing more than an argument against the institution of private property. The long and bitter experience with public ownership of resources in the former Soviet bloc suggests that the argument should be put the other way; lack of private ownership allows the exploitation of resources by some to the detriment of others. By contrast, a legitimate concern in the creation of an ITQ system is the mechanism used to distribute the initial quota rights. An auction favours those who have access to capital. A lottery favours those who are lucky. Allocation to existing fishermen favours history, and is politically the most feasible and most appropriate option from an economic perspective (Anderson and Libecap 2010; Anderson et al. 2011).

One additional criticism of ITQs is that such schemes are more expensive to administer and enforce than traditional types of schemes. All fisheries management schemes have costs, but the advantage of ITQs is that they focus attention on the explicit costs of management versus the economic benefits. Improvements to management are more likely to be initiated if the costs of management are transparent. As ITQs result in improved economic efficiency and profitable fisheries, they can pay for increased and more expensive monitoring and enforcement.[3]

ITQs in practice

Several countries have recognised the need for change and taken a different approach, after experimenting first with various regulatory regimes and witnessing their failure. Their emphasis is to rely more on managing the fisheries within a rights-based framework instead of management by direct control and regulation. Although no country has yet completely privatised their fisheries, many countries have experimented with property-rights-based management including Australia, Canada, the US, Chile, Peru, Namibia, South Africa, Norway, Russia, the Netherlands, the UK and several other European Union countries (Arnason and Gissurarson 1999; Shotton 2000; Arnason 2002; Hannesson 2004). In addition to these examples, New Zealand and Iceland have used property-rights management more

---

3 It seems that in countries that had adopted ITQs by 2000 the cost of enforcement was no greater and often smaller than in the other countries (Schrank et al. 2003). A suggested explanation may be found in the ineffectiveness of other types of management systems, that led governments to implement increasingly complicated and costly measures to address the resulting problems (see Arnason 2012).

extensively than other countries. Here we will look briefly at the Icelandic experience with fisheries management in recent decades (Runolfsson 1999; Arnason 1995, 2005).

Iceland was one of the first nations to adopt the ITQ system in its fisheries in the 1970s and 1980s, and there is therefore considerable evidence on the system's impact. Iceland is a moderately large fishing nation (19th on a global scale in 2009) and one of the most fisheries-dependent countries in the world. In recent years catches have amounted to about 1.6 million tonnes annually (reaching a peak of 2.2 million tonnes in 1997), some 2 per cent of the global marine harvest. About 40 per cent of its export earnings have been generated by fish products. The fishing industry has directly accounted for over 10 per cent of gross domestic product and, according to a recent estimate, directly and indirectly for up to 25 per cent (Arnason 2008).

Before the introduction of its ITQ system, Iceland experimented with a wide range of alternative fisheries-management systems. These included access licenses, fishing effort restrictions, investment controls and vessel buyback programmes, all of which were found to be unsatisfactory. The Icelandic ITQ system was created because of sharply declining stocks of herring in the late 1960s and early 1970s, and cod in the 1980s.

Following the extension of the exclusive fishing zone (EEZ) to 200 nautical miles, the major demersal fishery, the cod fishery, was subjected to an overall catch quota (TAC). The annual quotas recommended by the marine biologists soon proved difficult to maintain. Hence, individual effort restrictions, taking the form of limited allowable fishing days for each vessel, were introduced in 1977. The demersal fleet, however, continued to grow both through improvement of existing vessels and via new entry as it was still possible for new vessels to be added to the fleet. The annual allowable fishing days, therefore, had to be reduced from year to year. At the beginning of the individual effort restriction regime in 1977, deep-sea trawlers were allowed to pursue the cod fishery for 323 days only. Four years later, in 1981 this number of allowable fishing days for cod had been reduced to 215 days. This system was economically wasteful. Following a sharp drop in the demersal stock and catch levels, a system of individual vessel quotas was introduced in 1984.

Initially, the vessel quota system was instituted for one year only. Only vessels under 10 GRT (Gross Register Tonnage), which accounted for only a tiny portion of the demersal catch, were exempt from the quota system. In 1985, the system was extended for another year but with an important provision added; vessels preferring effort restrictions could opt for that arrangement instead of the individual quota restriction. This system was extended largely unchanged for an additional two years in 1986, and then again for the period 1988–90, and now included all vessels except small vessels using only hook and line gear. Although the acceptance of the individual vessel quota system was based on agreement between the government, parliament, the fishing industry and other stakeholders, policy challenges emerged in the late 1980s. For example, the catches of the many important species were still exceeding scientific advice and even the TACs decided by the government. The excessive fishing became unacceptable and there was substantial pressure to integrate different effort restrictions into a single management system so that all operators could use the same rules. These developments led to the Fisheries Management Act in 1990, providing a legal basis for a fairly uniform and comprehensive ITQ system.

This Act, which became effective in 1991 and is of indefinite duration, abolished the limited effort option in the demersal fisheries. Moreover, vessels between 6 and 10 GRT were incorporated into the ITQ system. However, the exemption from the ITQ system for vessels under 6 GRT was retained with the provision they could only use fishing gear based on 'hooks and line', i.e. fishing with any type of nets was forbidden. Since then, the ITQ system has been extended in several steps and now comprises practically all Icelandic fisheries.[4]

Before 1991, the ITQ systems in place were limited both in terms of the fisheries they applied to and fleet coverage. Several fisheries and fishing fleet classes were not covered and the continuation of the policy was somewhat uncertain. Long-term transfers of quota rights were still problematic and, as a result, quota holdings were generally

---

4   A comprehensive fisheries-management legislation stipulating ITQs as the main fisheries-management system was passed in 1990 (Act no. 38/1990). Since then, changes in the legislation and the associated regulations have been made almost every year. So many were the changes and amendments (35 by 2005) that the whole legislation was rewritten in 2006 (Act no. 116/2006). Since 2006 many further changes and amendments have been passed by the parliament.

not accepted directly as collateral by financial institutions. Fisheries management legislation had limited duration, for one to three years, due to the use of sunset clauses. The quality of the property rights created was therefore limited.

This changed with the 1990 Act, which made the ITQs indefinite. The system was formally established as the cornerstone of Icelandic fisheries management. Its coverage was greatly increased and its property-rights attributes were clarified. Thus, in spite of the small-vessel exemption (abolished in 2004), from 1991 onwards a high-quality ITQ system may be said to have applied in the Icelandic fisheries. However, the legislation for the system still does not establish perfect (harvesting) property rights. Most importantly, there is still some uncertainty about the system's permanence as a parliamentary majority could always revoke the legislation and revert to regulated open access. In addition, the quotas are subject to special taxation, which reduces the value of the property right.

The basic property right in the system is a share of the TAC for every species for which there is a TAC. The quotas are permanent (of indefinite duration), perfectly divisible and transferable.[5] The legislation has a provision for a ceil-

---

5 The term or duration of the TAC-shares is not stipulated in law. Although it is clear that they are not explicitly in perpetuity, they may turn out to be so. More precisely, according to legal opinion, the ITQ system may be abolished and the TAC-shares withdrawn without compensation to the holders, provided a notice of several years is given. Therefore, this basic asset of the ITQ system must be regarded as being of uncertain duration. TAC-shares, however, are secure in the sense of being protected by law like any other asset and they exhibit certainty in exclusivity over the corresponding harvests.

ing or maximum quota holding for individual species as well as an overall ceiling for all species.[6] The permanent TAC-shares held by any company or individual are subject to an upper bound that ranges from 12 per cent of the TAC for cod up to 35 per cent of the TAC for ocean redfish. Moreover, an individual company must not control more than 12 per cent of the value of all TACs. These stipulations are explicit to prevent what is regarded by parliament as excessive concentration in the fishing industry.

The cost of administering and monitoring the ITQ system in Iceland has not been greater than expected. The Fisheries Management legislation indirectly provides for cost recovery of fishery management costs. In addition to a (small) fee for commercial fishing licenses, there is both a general tax and a special tax on quotas, and the former may be seen as a cost recovery measure. The Icelandic government operates the Marine Research Institute (MRI), which conducts oceanographic and fisheries research and makes recommendations about annual TACs in different species of fish to the Ministry. Its operating costs are paid out of the government budget.

What has the ITQ system in the Icelandic fisheries achieved and what could it be expected to achieve? Some

---

6   ITQs or TAC-shares are calculated in so-called cod equivalents. The term 'cod equivalent' refers to weight and implies the relative value of different fish species on the market compared with cod. For each vessel having a quota for several species, the total quota may be calculated in kilos as cod equivalents. This aggregate value is calculated in cod equivalents using species exchange rates (essentially price ratios) set annually by the Department of Fisheries and Aquaculture within the Ministry of Industries and Innovation.

critics have claimed that it has not resulted in a recovery of fish stocks, cod stocks in particular. As cod stocks are recovering and stocks in general are in a stable condition, this criticism is misplaced. Resource conservation is achieved by setting the total quota appropriately, no matter what system of fisheries management is adopted. The ITQ system is mainly a tool to achieve economic efficiency, given that a TAC and ITQs also help conservation by making it easier to keep the catch within the set limits and by fostering an attitude of conservation among quota holders.[7] The total catch quotas in the Icelandic fisheries were simply set too high by the government during most of the past few decades and it is only in recent years, with support and even effective pressure from the industry, that TACs have been more conservative. Setting catches more conservatively is in the long-term interest of the industry when they have a stake in increased future catches.[8]

The experience with the ITQ system is generally favourable. The Icelandic summer-spawning herring stock was

7   A survey by Branch (2008) of more than 200 peer-reviewed papers on the effects of ITQ programmes reports that participants in catch-share fisheries often support lower TACs. Based on this, there seems to be a general tendency that the adoption of catch-share reforms encourages fishers to support lower and more sustainable TAC limits.

8   The government therefore now follows the TAC advice of the MRI very closely. The recommended TAC by the MRI, it should be noted, is the biologically based maximum sustainable yield (MSY) and not (necessarily) the economically MSY. The difference in essence is that the MRI is trying to maximise the biological yield (the maximum catch from a sustainable stock) whereas an economist (or owner) would maximise the economic yield; the long-term profit (maximum catch for maximum sustainable profits).

the first fishery where ITQs were initiated, when the fishery was reopened in 1975 after it collapsed in the late 1960s. Catches of herring increased and, more importantly, catch per unit effort has increased significantly.[9] The number of vessels in the fishery has declined from more than 200 in 1980 to fewer than 30 by 1995, although the average vessel size has increased substantially.

The demersal fisheries, for example, cod fisheries, have been slow to improve, one reason being that the TACs were set too high in the 1980s and there was still fishing in excess of TACs into the 1990s. Politicians chose a gradual approach to reducing the cod catch, despite recommendations by the Marine Research Institute (MRI) for steeper cuts. Only more recently has the TAC been close to the levels suggested by the MRI. That was done at the insistence of the Association of Vessel Owners (an organisation of the owners of larger vessels), which wants to preserve the value of their ITQ assets. Stocks seem to have rebounded in recent years. Both the fishable stock and the spawning stock of cod have grown over the last few years and the spawning stock is now more than twice as large as it was for most of the last decade. Indeed, it hasn't been this big since the early 1960s. The fishing mortality rate of cod has decreased and the harvest rate (proportion of the fishable stock) has also decreased. This change means that year classes last longer in the overall population and stocks are growing as a result. The proportion of older fish

9   In 2009 the Icelandic summer-spawning herring stock was heavily infected by *Ichthyophonus*. It is estimated that roughly 40 per cent of the stock died because of the infection but it has slowly recovered since then.

in catches has increased despite the fact that rather small year classes are now the majority of the fishable stock. These effects are seen in increased catch per unit effort and more economical use of allowed quotas.

As noted above, high grading – the discarding of less valuable catch – is a problem often attributed to ITQ systems, especially in mixed fisheries. The Icelandic demersal fisheries are certainly mixed fisheries. Nevertheless, there is little evidence of increased discarding under the ITQ system. In fact, according to measurements by the MRI, discards in the most important demersal species are only 1 per cent of average of total catch volume.[10] The rate of discards has also declined since the introduction of ITQs. Discards in pelagic fisheries are also insignificant.

As mentioned above, small vessels were initially exempted from the ITQ system, with the aim of protecting 'little' fishermen from 'big' fishermen. Predictably, this exemption resulted in a large increase in the number of small vessels. To counter this increase, several measures were introduced. In 1988 small vessels were limited to only using fishing gear based on 'hooks and line' and in 1991 vessels between 6 and 10 GRT were incorporated into the ITQ system. Finally, in 2004 a separate ITQ system for the fleet of small vessels with hook-and-line permits was put

---

10 Discards depend on gear and vessel type and can amount to a high of almost 5 per cent, although the total average is less than 1 per cent (see http://www.hafro.is/Bokasafn/Timarit/fjolrit-171.pdf). There is some flexibility in the Icelandic system (see above), such as allowing catch in one species to be covered by quota of another species, with 'cod equivalent rates' and by allowing landings of some amount of juvenile fish that is not counted towards the fisher's quota.

in place.[11] About 300 small vessels were active in fishing in 1984 and this had increased to more than 2,000 in the early 1990s, but in 2012 there were only 342 in the small vessel ITQ system.[12]

Since 1990, when the comprehensive ITQ system went into effect, there have been substantial improvements in the economic efficiency of the demersal fisheries. Total fishing effort went down by more than 30 per cent in the first 10–15 years after the ITQs were introduced. Fishing capital, which had increased by more than 400 per cent during the period 1960–90, has actually declined since 1990, and the number of vessels has also declined. In 1992/93, there were 1,265 vessels with ITQs and another 1,125 with hook licenses, or 2,390 in total (there were some 162 additional vessels with commercial fishing licenses but without quota). In September 2012 only 603 vessels in total were allocated quotas (had ITQ shares), of which 261 were in the ITQ system for larger vessels and 342 in the small-vessel ITQ section (the total number of fishing vessels in Iceland was 1690 in January 2012). This reduction in the number of vessels, and increased quota concentration

---

11 With the change in 1988 a number of small vessels chose to receive quotas and become part of the ITQ system so that they could continue to use the fishing gear of their choice. Their individual quota was based on their catch in previous years. This provided an incentive for other small vessels to race for quota and not only race for fish. That is, they invested in a catch record, expecting this would determine their individual quota in the future, when all commercial vessels would be incorporated into ITQs.

12 As part of this process of small vessels being subject to an ITQ system the vessel size limits have been changed from the initial 6 GRT in 1991 to 15 GRT in 2008, and 25 GRT in 2014.

Figure 4 Profits in the Icelandic fisheries industry, 1980–2012

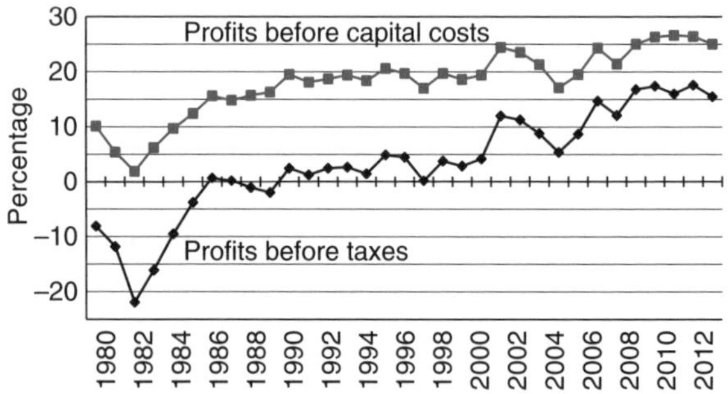

Note: Net profit as a % of revenue before and after (imputed) cost of capital, based on actual accounts.
Source: Statistics Iceland.

at the same time, is financed by the fishing industry itself. That is, the fishing firms buy each other out and improve their efficiency, without the state being directly involved or government subsidies.

The main purpose of the ITQ system is to improve the economic efficiency of the fisheries. The Icelandic fisheries are biologically very productive and should be able to generate high economic rents. Until the adoption of the vessel quota system, however, comparatively low rents were generated in the industry. In fact, during the years preceding the introduction of the vessel quota system, industry profits were often highly negative (see Figure 4). Since the introduction of ITQs the quality of the harvest and profits have improved significantly and, as mentioned above, fishing effort has been reduced. Overall productivity and efficiency has therefore increased greatly.

Conclusion

The current global marine catch could be harvested with approximately half of the current global fishing effort. In other words, there is massive overcapacity in the global fleet. Excess fleets competing for limited fish resources result in stagnant productivity and economic inefficiency. In response to the decline in physical productivity, the fishing industry has attempted to maintain profitability by reducing labour costs, lobbying for subsidies and increasing investment in technology. Partly as a result of its poor economic performance, real income levels of fishers remain depressed as the costs per unit of harvest have increased.

From an economic perspective the race to fish, the drive to increase fishing power, and the perversion of the politics of the management process are all driven by the insecurity of access faced by fishermen under most fisheries-management systems. Insecure harvest rights provide distorted incentives and lead fishermen to compete wastefully with each other and with fisheries managers.

Rights-based systems have dramatically changed individual incentives and the behaviour of fishermen in fundamental ways. This change in behaviour is broad-based and persistent, and arises because security of access allows fishermen to shift attention away from attempting to capture larger shares of a fixed pie and towards maximising the value from the secure shares that they command under rights-based systems. This change has profound effects on all dimensions of fishing, from harvesting strategies, to investment, to stewardship of the resource, to marketing innovations, to conducting science and fish stock assessment. By contrast, traditional top-down management systems, with their input and output control methods, fail to generate long-term stewardship incentives and therefore perpetuate the adversarial relationship between users and regulators.

Although the theoretical shortcomings of institutions based on property rights have been argued about for years, there is now enough evidence to enable a focus on empirical results rather than mere speculation, theoretical or otherwise. Almost all the relevant experience suggests that rights-based management institutions alter incentives in ways favourable to conservation and stewardship. A very important inducement for behavioural changes is the wealth capitalised in the value of tradable quotas in such systems.

The rapid adoption of ITQ systems around the world is indicative of their relative success in overcoming the commons problem and improving the economics of fisheries. Empirical evidence confirms that ITQs have reduced excessive fishing effort and overcapitalisation in fisheries and significantly increased the unit value of landings. These improved economic results reflect improved allocative efficiency, which is a virtually inevitable outcome of any reasonably designed and enforced ITQ system.

References

Anderson, T. L. and Libecap, G. D. (2010) The allocation and dissipation of resource rents: implications for fishery reform. In *Political Economy of Natural Resource Use: Lessons for Fisheries Reform* (ed. D. Leal), pp. 79–95. World Bank.

Anderson, T., Arnason, R. and Libecap, G. D. (2011) Efficiency advantages of grandfathering in rights-based fisheries management. *Annual Review of Resource Economics* 3(1): 159–79.

Arnason, R. (1995) *The Icelandic Fisheries: Evolution and Management of a Fishing Industry.* Oxford: Fishing News Books.

Arnason, R. (2002) A review of international experiences with ITQs. CEMARE, Report 59, University of Portsmouth.

Arnason, R. (2005) Property rights in fisheries: Iceland's experience with ITQs. *Reviews in Fish Biology and Fisheries* 15(3): 243–64.

Arnason, R. (2008) Iceland's ITQ system creates new wealth. *The Electronic Journal of Sustainable Development* 1(2): 35–41.

Arnason, R. (2012) Property rights in fisheries: how much can Individual Transferable Quotas accomplish? *Review of Environmental Economics and Policy* 6(2): 217–36.

Arnason, R. and Gissurarson, H. (eds) (1999) *Individual Transferable Quotas in Theory and Practice.* Reykjavik: University of Iceland Press.

Arnason, R. and Runolfsson, B. (eds) (2008) *Advances in Rights Based Fishing; Extending the Role of Property in Fisheries Management.* Reykjavik: Ugla publishing and RSE (Centre for Social and Economic Research).

Branch, T. (2008) How do Individual Transferable Quotas affect marine ecosystems? *Fish and Fisheries* 10.

Costello, C., Gaines, S. D. and Lynham, J. (2008) Can catch shares prevent fisheries collapse? *Science* 321: 1678–81.

Costello, C., Lynham, J., Lester, S. E. and Gaines, S. D. (2010) Economic incentives and global fisheries sustainability. *Annual review of resource economics* 2(1): 299–318.

Grimm, D., Barkhorn, I., Festa, D., Bonzon, K., Boomhower, J., Hovland, V. and Blau, J. (2012) Assessing catch shares' effects evidence from Federal United States and associated British Columbian fisheries. *Marine Policy* 36: 644–57.

Hannesson, R. (2004) *Privatization of the Oceans*. Cambridge, MA: MIT Press.

Neher, P., Arnason, R. and Mollet, N. (1989) *Rights-based Fishing*. Boston: Kluwer Academic Press.

Runolfsson, B. (1999) The Icelandic system of ITQs: its nature and performance. In *Individual Transferable Quotas in Theory and Practice* (ed. R. Arnason and H. Gissurarson). Reykjavik: University of Iceland Press.

Schrank, W. E., Arnason, R. and Hannesson, R. (2003) *The Cost of Fisheries Management*. Aldershot: Ashgate.

Scott, A. D. (2008) *The Evolution of Resource Property Rights*. Oxford University Press.

Shotton, R. (ed.) (2000) Use of property rights in fisheries management. FAO Fisheries Technical Paper 404/1 & 2. Rome: FAO.

Wilen, J. E. (2006) Why fisheries management fails; treating symptoms rather than the cause. *Bulletin of Marine Science* 78: 529–46.

Wilen J. E., Cancino, J. and Uchida, H. (2012) The economics of territorial use rights fisheries, or TURFs. *Review of Environmental Economics and Policy* 6(2): 237–57.

World Bank and FAO (2009) *The Sunken Billions: The Economic Justification for Fisheries Reform*. Washington, DC: World Bank/FAO.

ABOUT THE IEA

The Institute is a research and educational charity (No. CC 235 351), limited by guarantee. Its mission is to improve understanding of the fundamental institutions of a free society by analysing and expounding the role of markets in solving economic and social problems.

The IEA achieves its mission by:

- a high-quality publishing programme
- conferences, seminars, lectures and other events
- outreach to school and college students
- brokering media introductions and appearances

The IEA, which was established in 1955 by the late Sir Antony Fisher, is an educational charity, not a political organisation. It is independent of any political party or group and does not carry on activities intended to affect support for any political party or candidate in any election or referendum, or at any other time. It is financed by sales of publications, conference fees and voluntary donations.

In addition to its main series of publications the IEA also publishes a quarterly journal, *Economic Affairs*.

The IEA is aided in its work by a distinguished international Academic Advisory Council and an eminent panel of Honorary Fellows. Together with other academics, they review prospective IEA publications, their comments being passed on anonymously to authors. All IEA papers are therefore subject to the same rigorous independent refereeing process as used by leading academic journals.

IEA publications enjoy widespread classroom use and course adoptions in schools and universities. They are also sold throughout the world and often translated/reprinted.

Since 1974 the IEA has helped to create a worldwide network of 100 similar institutions in over 70 countries. They are all independent but share the IEA's mission.

Views expressed in the IEA's publications are those of the authors, not those of the Institute (which has no corporate view), its Managing Trustees, Academic Advisory Council members or senior staff.

Members of the Institute's Academic Advisory Council, Honorary Fellows, Trustees and Staff are listed on the following page.

The Institute gratefully acknowledges financial support for its publications programme and other work from a generous benefaction by the late Professor Ronald Coase.

Other IEA publications

Comprehensive information on other publications and the wider work of the IEA can be found at www.iea.org.uk. To order any publication please see below.

Personal customers

Orders from personal customers should be directed to the IEA:

Clare Rusbridge
IEA
2 Lord North Street
FREEPOST LON10168
London SW1P 3YZ
Tel: 020 7799 8907. Fax: 020 7799 2137
Email: sales@iea.org.uk

Trade customers

All orders from the book trade should be directed to the IEA's distributor:

NBN International (IEA Orders)
Orders Dept.
NBN International
10 Thornbury Road
Plymouth PL6 7PP
Tel: 01752 202301, Fax: 01752 202333
Email: orders@nbninternational.com

IEA subscriptions

The IEA also offers a subscription service to its publications. For a single annual payment (currently £42.00 in the UK), subscribers receive every monograph the IEA publishes. For more information please contact:

Clare Rusbridge
Subscriptions
IEA
2 Lord North Street
FREEPOST LON10168
London SW1P 3YZ
Tel: 020 7799 8907, Fax: 020 7799 2137
Email: crusbridge@iea.org.uk